TRAWLING IN THE MIST:

INDUSTRIAL FISHERIES IN THE RUSSIAN PART OF THE BERING SEA

by Alexey Vaisman

Funding for the production of this report was kindly provided by WWF-US and the David and Lucile Packard Foundation

Credit: Alexey Vaisman, TRAFFIC Europe-Russia

Fish in the hold of a trawler impounded in the Russian part of the Bering Sea

TABLE OF CONTENTS

ACKNOWLEDGEMENTS

The kind support of WWF-US allowed this report, and the investigations on which it is based, to be translated into reality. TRAFFIC is also grateful to the David and Lucile Packard Foundation for assisting with funding for production of the report.

During field work and collection of information, investigators encountered people with an understanding of the importance of the work they did. I extend thanks to staff of agencies, those aboard vessels, fisheries inspectors, journalists and staff of research institutes for their help and information. Many of them helped with an air of willingness and friendship. Special thanks go to the crew of the inspection ship *Ingeneer Martynov*.

Really great help and moral support were rendered by the then Head of Kamchatka Regional Directorate for Protection and Reproduction of Fish Stocks and Regulation of Fishery (*Kamchatrybvod*), Vladimir Burkanov.

Reviewing the report entailed the efforts of several people. I want to thank Margaret Williams and Tom Grasso of WWF-US and their colleagues and Nina Marshall and Stephanie Theile from TRAFFIC Europe, who put much effort into adapting the text for readers outside the Russian Federation. Anna Willock from TRAFFIC Oceania, Clarence Pautzke of the North Pacific Fishery Management Council, Alexander Shestakov and Vassily Spiridonov from the WWF Russia Programme Office have also devoted valuable effort to improving the report.

Especial thanks go to Caroline Raymakers from TRAFFIC Europe. Her help in compilation of statistics was really fantastic! She also played a great role in the process of structuring the report. I felt Caroline's friendly support and help all the while I worked on this project! To be honest, in my heart I believe Caroline Raymakers to be a co-author of this report.

Julie Gray from TRAFFIC International is an editor of this issue. She courageously straightened out the cobwebs of the Bering Sea fishery and trade in fish products. Thanks a lot for nice work.

Alexey Vaisman
TRAFFIC Europe, Russia Office
WWF Russia Programme Office
P O Box 3
109240 Moscow
Russian Federation

EXECUTIVE SUMMARY

Background

The Bering Sea separates some of the most northerly regions of the Russian Federation and the USA, stretching northwards from the Aleutian Islands as far as the Arctic Circle. Local human populations have lived off the vast diversity of fish, marine mammals and other life forms of these waters for centuries. During the twentieth century, however, the rate of harvest increased to the point where regeneration for some species was affected. For others, signs of over-fishing are less clear, yet what is certain is that effective management of fisheries in the Russian part of the Bering Sea has largely collapsed during the 1990s.

Introduction

Between them, the Russian Federation and the USA encompass 92% of the Bering Sea within their territorial waters and EEZs (Exclusive Economic Zones). In terms of national jurisdiction, regulation of fishing in much of the Bering Sea falls to these two countries, therefore. Internationally, however, efforts are underway to influence conservation of the Bering Sea and its life systems. With this aim, the Bering Sea has been designated a global priority area for conservation - a so-called ecoregion - by some environmental ngos, including WWF and TRAFFIC. While it is understood that conservation of this region is a priority, the most urgent conservation needs within the area are still emerging as more information is gathered. TRAFFIC carried out this investigation into the industrial fishing of the Russian part of the Bering Sea with the aims of exploring:

- the evolution of commercial fisheries in the region;
- the legislative and enforcement structure governing fisheries;
- key target fisheries, including catch and trade levels over time;
- illegal practice and factors conducive to this.

Methods

The project was undertaken by TRAFFIC Europe - Russia Office, and research in the field was based in Kamchatka, one of the easternmost provinces of the Russian Federation, with a prominent fishing industry. As such, it was selected as representative of the industry in the Russian part of the Bering Sea as a whole. Investigations in Kamchatka took place from July to December 1999 and included a two-week trip on board an enforcement patrol vessel. Observations and interviews on board the patrol ship and on land in Kamchatka focused on collecting information on fisheries management, trade, legislation and enforcement, and illegal activities. Desk-based research on the project continued into 2001 and centred on accessing relevant literature and catch and trade statistics and verifying, interpreting and supplementing findings from the field.

Overview

Industrial-scale fishing in the Russian part of the Bering Sea with an external trade focus has its beginnings in the late nineteenth century. The 1950s were a time of expansion and development of the fishing sector, which by the 1980s had become the industry on which the economy of Kamchatka depended. With State support and regulation, fishing provided a stable form of employment, but as such assistance and control were withdrawn, as the political character of the government changed, the industry lost its *modus operandi* and funding base. Partnerships with foreign companies - joint ventures - proliferated and other enterprises were undertaken in attempts to harness new sources of profit from fishing, chiefly from hard currency earned by selling abroad. Despite such initiatives, the fishing industry of Kamchatka found it impossible to adapt itself within a decade to the changed demands required of it and at the end of the 1990s, fishing was a loss-making industry in Kamchatka.

Administration of the fisheries industry

Fisheries in the Russian Federation are administerd at federal level by the State Committee of Fisheries, which is assisted by scientific advisory bodies and at regional level by branch directorates (*rybvods*). The State Committee is responsible for formulating and governing all aspects of policy and management of the Russian fishing industry, from scientific research, to determining legal gears, zones and seasons, to marine safety and socio-economic considerations. At present, the national legislation relevant to commercial fishing in the Russian Federation, dating from 1989, is largely outdated and currently under review. A number of international agreements have relevance for the regulation of fishing in the western Bering Sea.

The general fishing quota within a given region in divided into four categories, including an allocation to be distributed free to Russian companies, a quota intended for research purposes and quotas for sale. The biological criteria by which catch quotas are set are unclear: quotas for Alaska Pollack *Theragra chalcogramma* in the Russian part of the Bering Sea have risen between 1996 and 1999, yet stocks of the species have been in decline. Disputes over quota allocations have been fiercely competitive, resulting in human deaths in at least one case, in 1999.

Catch records for the Russian part of the Bering Sea are not readily available for recent years (since 1993) but catch statistics from a wider area of the north-west Pacific, including the western Bering Sea, were accessed for contextual purposes. For the period 1984-99, a decline can be noted in Russian catches of Alaska Pollack from this area and, to a lesser extent, in those of Pacific Cod *Gadus macrocephalus*, while Pacific Herring *Clupea pallisii pallisii* catches have increased. References to catches from the north-east Pacific area, which includes the eastern Bering Sea, are made in the spirit of ecoregional conservation, as a means of keeping in mind that pressure on stocks is not limited within any geographical or political boundary.

Responsibility for enforcement of Russian fisheries legislation in the Bering Sea is shared between several State agencies. Foreign vessels are required to have an inspector permanently on board, while other vessels are monitored by roving patrol vessels and from the air, as well as at specific checkpoints. Exceptional enforcement missions, or *putina*, are also undertaken, on a joint basis, by the various enforcement agencies.

Target fisheries

The main commercial marine resources targeted by industrial fisheries in the Russian part of the Bering Sea are Alaska Pollack; Pacific Cod and Saffron Cod *Eleginus gracilis*; several species of crab; Pacific Herring; rockfishes *Sebastes* and *Sebastolobus* spp.; various halibuts and flounders *Hippoglossus stenolepis* and *Reinhardtius* spp.; plaices, especially Yellowfin Sole *Limanda aspera*; the Commander Squid *Berryteuthis magister magister* and shrimps.

Accounting for half the biomass of groundfish in the Bering Sea, and as the basis of one of the largest single-species fisheries in the world, Alaska Pollack is the most important species to the fishing industry of the Russian Bering Sea. The biomass of Alaska Pollack in the western Bering Sea has declined markedly between 1991 and 1996, according to scientists based at the Pacific Research Institute of Fisheries and Oceanography (TINRO). Despite this, from 1996 to 1999 catch quotas for the area rose and exports from Kamchatka have increased over more or less the same time period. The quest for pollack has been keen, especially in the light of demand for new pollack products, such as roe, fillet and minced fish to make *surimi* (for the manufacture of "crab" sticks and other processed seafood products). The outlook for pollack fishing in the Russian part of the Bering Sea appears unfavourable, especially since illegal catches are estimated to equal half the legal quota for the species, according to staff of *Kamchatrybvod* (the Kamchatka Regional Directorate for Protection and Reproduction of Fish Stocks and Regulation of Fishing).

Cod fishing has consistently been the second biggest fishery in terms of volumes caught (1980-93) in the Russian part of the Bering Sea, but for this and several other species reviewed, trends in biomass levels, catches and trade are hard to quantify. Interest in the rockfish fishery has reportedly been lessening among Russian fishers and some declines in catch levels perceived, while crab quotas were reduced several-fold between 1997 and 1998, suggesting reduced supply. By contrast, herring stocks are believed to be increasing and squid stocks are reported not to be fully exploited. Catches and exports of squid may increase in coming years, particularly if other stocks dwindle.

Exports

Asia is the prominent destination for exports of marine products from Kamchatka. The Republic of Korea, Japan and China were the most significant importers from 1995-98. During the second half of the 1990s, the overall volume of exports rose, with a particularly noticeable leap from 1996 to 1997. Since 1997, Alaska Pollack has been the dominant export from

Kamchatka in terms of volume. Crab exports, which had been the most valuable fisheries export from Kamchatka since 1995, were outstripped in value by pollack exports in 1998 as volumes of these continued to rise.

Violations of fisheries legislation

Investigators found evidence of illegal activities at virtually all levels of the fishing industry, while in Kamchatka. Illegal practice in the sector proliferated during the 1990s. As the dependable State income and subsidies to those in the fishing industry dwindled in the early part of the decade, the need to find alternative sources of finance provided an incentive to make money through unauthorized means. Estimates of the value of trade lost to the Russian Government through illegal exports of fisheries products are put at one to five billion US dollars.

Types of violation are numerous. During the period 1993-98, the number of Russian vessels committing infractions was consistently more than double the number of non-Russian vessels offending and, at its peak, the ratio was 51:1, respectively. The most common form of offence was found to be use of fraudulent documents. Falsifying information in documents was often found to be the first step in facilitating a string of other violations, notably exceeding catch quota limits, unauthorized sale of over-quota harvest and undocumented export of the same. Listed in order of descending frequency of incidence, re-sale of quotas; fishing in prohibited areas; concealment of prohibited types of catch; use of banned fishing gear; unauthorized processing of catch, usually of crabs or salmon on board vessels; pollution of the sea and fishing without a licence were other illegal activities encountered or reported.

Factors contributing to illegal activities

Besides the initial motivation to supplement or earn income, there are several factors conducive to illegal fishing-related activites in the Russian part of the Bering Sea.

Inappropriate legislation

The legislative framework for controlling much illegal activity is either missing or out-of-date and as a result unclear or inapplicable. For example, according to current regulations, exports of marine bioresources harvested in the Russian EEZ, but outside its territorial waters, are not subject to normal Customs clearance procedures, provided they are sold without entering Russian territorial waters. Penalties for these and other offences, on the other hand, are arguably too low to act as a deterrent.

Weaknesses in the enforcement system

An effective system for enforcement of any legal framework to control of fisheries in the Russian part of the Bering Sea is also missing. Poor co-ordination between enforcement agencies is acknowledged and the problems this causes are compounded by inadequate enforcement equipment, including vessels, low-paid staff and corruption within the network of observers on board foreign ships.

Organized crime

The existence of organized criminal operations in the Bering Sea fisheries was evident during investigations for this report. Criminal organizations plan and oversee illegal fishing, sometimes reportedly issuing a document known as a "provisional instruction" to guide captains through procedures to minimize the risk of being caught by law enforcement agents. Sometimes groups of vessels work together, using radio communications, to increase the chances of evading inspection, one ship acting as a kind of look-out for others fishing illegally.

Conclusion

An unavoidable conclusion of investigations into the fisheries of the Russian part of the Bering Sea is that regulation of the industry is in disarray. Recent, accurate, baseline information on fish stocks and catches is often uncertain or unavailable, with the consequence that fisheries management is lacking a sound foundation for decision-making. Instead of adopting a precautionary approach in the circumstances, however, Russian authorities have increased quotas for some fish without an obvious biological basis for so doing. The potential seriousness of incautious management is emphasized when one reflects that the Bering Sea is of critical importance to Russian fisheries, providing around half of the country's marine harvest in the 1990s. Recorded fisheries exports from Kamchatka have been increasing in volume, from 1995 to 1998, but the unit value of several of these commodities has dropped over this period. This, and the fact that a large proportion of fish is caught and sold without passing through State-approved channels means that little income from fisheries is being harnessed by the government for re-investment in the industry. Legal and enforcement mechanisms in the Russian Federation are incapable of redressing the loss of billions of US dollars to illegal operators annually. The solution is not only to strengthen these mechanisms, however, but to engage action on an international scale for a marine environment of special significance for the world.

Recommendations

As this report addresses industrial fisheries, these recommendations should, for the most part, be considered in relation to the large fleets of the industrial fishery of the Russian part of the Bering Sea.

Fisheries management

The Russian Government should take action to ensure:

- improved fisheries information, including species-specific surveys of fish stocks and the transmission of up-to-date catch data to the UN Food and Agriculture Organization (FAO)
- improved management of stocks, in such a way that a precautionary approach to the

management of industrial fisheries in the Bering Sea is adopted and the criteria for quota allocations are made transparent to stakeholders

- the identification and creation of protected areas in key habitats for important fishery stocks
- improved regulation of fishing gears, specifically by extending regulations to require the prohibition of all non-selective and destructive gear
- that governance over fishing and trade in the Russian EEZ of the Bering Sea is strengthened
- that social and community considerations are addressed, by requiring that people living adjacent to the Bering Sea be involved in decision-making affecting the resources on which they rely, and that their economic and community interests be balanced against the needs of industrial fisheries
- improved financing, through channelling fines for fisheries infractions, money from quota sales and other forms of fisheries income into reforms necessary in the fishing industry

Enforcement

The Russian Government should ensure that fisheries law enforcement is strengthened by:

- clarifying roles of, and improving co-ordination between, enforcement agencies
- improving the system of observers, by creating and co-ordinating a network of observers with new operating conditions, to reduce opportunites for corruption inherent in the current system
- expanding observer coverage to include Russian vessels and possibly to Customs duties, where applicable
- improving equipment, including satellite vessel monitoring systems
- adjusting financial incentives and disincentives, including increasing penalties and considering a bonus system for enforcement staff

International co-operation

At an international level, actions should be taken so that:

- interaction between Customs agencies of countries trading in Bering Sea marine resources is improved
- all nations involved in trade in Bering Sea fishery products apply the most precise category code available of the Harmonized Commodity Description and Coding System
- the implications of closure of the Donut Hole to Alaska Pollack fishing on marine resources in the western Bering Sea are examined under the *Convention on the Conservation and Management of Pollack Resources in the Central Bering Sea*
- the importance of bilateral US-Russian decision-making is emphasized

Awareness

Actions to increase awareness of the issues surrounding the conservation of marine resources in the western Bering Sea should include:

- a conference bringing together industry, regulatory agencies and environmental groups
- dissemination of information on the levels of threat to fish stocks to interested parties, with the aim of involving non-governmental groups, including industry, in funding or lobbying
- consideration of the use of economic incentives for the promotion of sustainable fisheries through certification or other trade mechanisms

BACKGROUND

The Bering Sea is the greatest sea basin in the northern Pacific Ocean, covering about a million square miles (2 269 000 km^2) and descending to a depth of 5121 m (Anon., 1998a). Named after Vitus Bering (1681-1741), a Danish-born officer of the Russian navy, it lies between Siberia and Alaska, extending from 155° E to 155° W, and from 50° N to the Arctic Circle. At the narrowest part of the Bering Sea, the Bering Strait separates the Russian Federation and the USA, the boundary lying between Big and Little Diomede Islands. Biologically, it is one of the most productive seas in the world, with a diverse and rich supply of fauna and flora. Over 450 species of fish and shellfish occur in the Bering Sea, as do 50 species of seabirds and 26 species of marine mammals (Banks *et al.*, 1999), among them commercially valuable resources - and all form part of a unique ecosystem, the linkages and complexities of which form the basis on which current and future resource exploitation depends.

Bering Sea resources have sustained the indigenous peoples of the region for thousands of years, but during the twentieth century, exploitation of the Sea's biological wealth has entered a different league. It is already known that the Sea has suffered declines in fish and crab numbers and in populations of some mammals and birds, such as Steller Sea Lions *Eumetopias jubatus*, Northern Fur Seals *Callorhinus ursinus,* Spectacled Eiders *Somateria fischeri* and the Red-legged Kittiwake *Rissa brevirostris,* which breeds only in the region. Such depletions and other forms of threat to the Bering Sea ecosystem have come from commercial harvest from the sea, the introduction of non-native species, coastal mining, pollution from shipping and shrinking ice cover (Banks *et al.*, 1999; Cline and Williams, 2000).

Between them, the Russian Federation and the USA have about 92% of the Bering Sea within their Exclusive Economic Zones (EEZs). For both these nations, the Bering Sea is of paramount importance. An estimated 50% of the Russian Federation's annual fish production and over half of the USA's annual fish catch came from the Bering Sea during the 1990s (Cline and Williams, 2000). Exploitation of the resources of the Bering Sea is also undertaken for markets far beyond local shores. In line with a global trend of increased pressure exerted on the marine biosphere, exploitation of the resources of the Bering Sea has become increasingly competitive and, by some accounts, fishing in the Russian part of the Bering Sea has become, a "free for all" for fishing fleets, both Russian and foreign (Bakilin and Ignatenko, 1996). Catches of some of the most important commercial marine species have diminished, although overall catch levels have remained relatively stable as a wider variety of species has been targeted (Anon., 1999a). Such a trend, where "increases in global landings.....are largely counteracted by declines in individual resources, reflecting the fact that a significant number of stocks have begun to be overexploited over at least the last decade" has been observed on a global scale (Anon., 1994a).

Set against such a background, management and regulation of the resources of the Bering Sea are clearly essential, but such control is often difficult when stewardship is shared between nations, and exploitation undertaken by many. These conditions present a challenge for management of many marine resources, but management in the Bering Sea is complicated by poor socio-economic conditions in the Russian Federation, yet at the same time particularly

Figure 1

Map to show the Bering Sea, including the Bering Sea ecoregion and other places referred to in this report

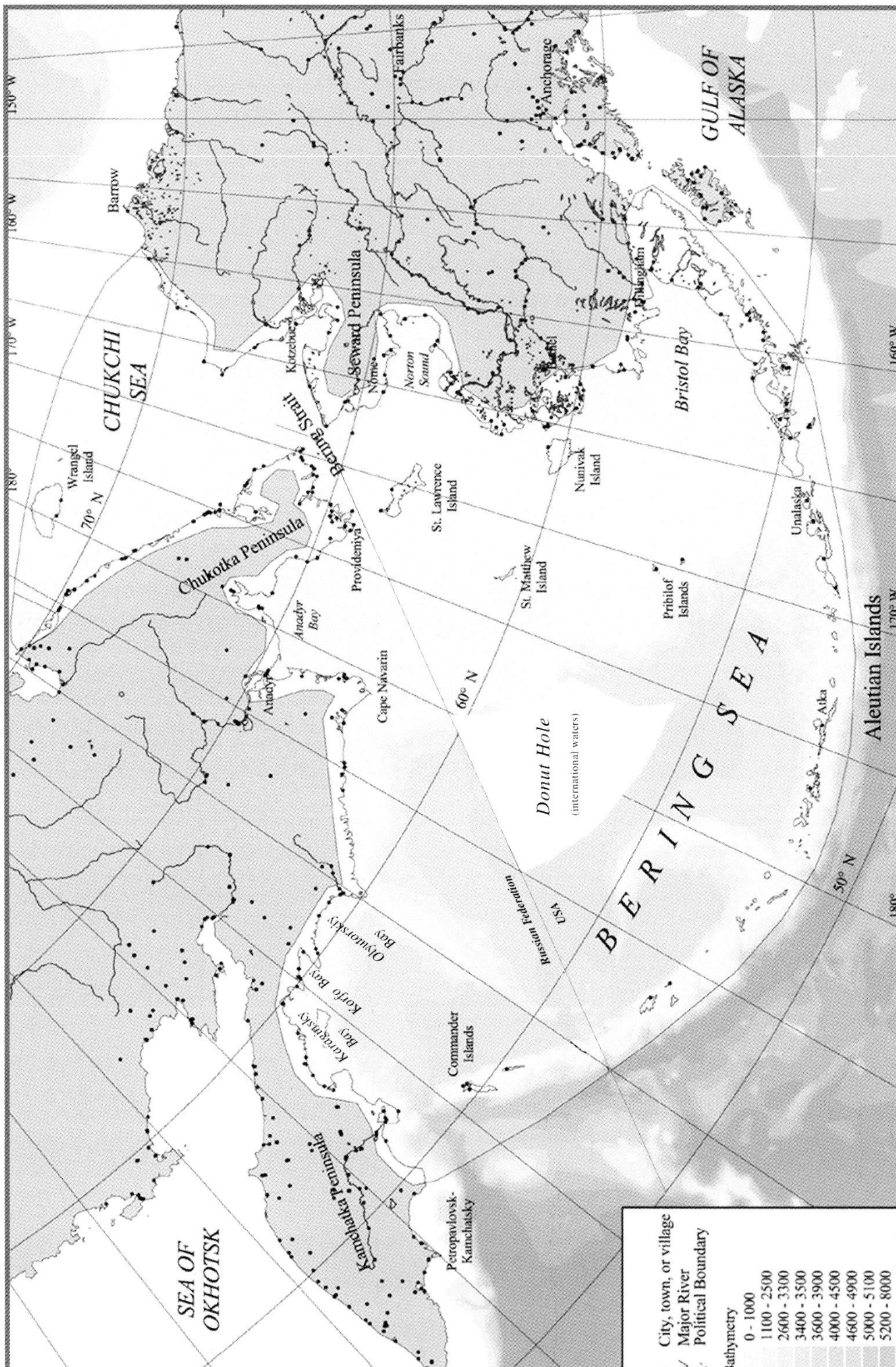

Credit: Adapted from a map prepared by Emma Underwood and Jennifer D'Amico, Conservation Science Program, WWF, May 1999. From: Banks, David, Williams, M., Pearce, J., Springer, J., Hagenstein, R., and Olson, D., 1999. *Ecoregion-based conservation in the Bering Sea: identifying important areas for biodiversity conservation.* World Wildlife Fund and The Nature Conservancy of Alaska, USA.

Scale = 1:10 446 281 m. Bathmetry measured in metres. ◡ = edge of Bering Sea Ecoregion

Note: Donut Hole as shown represents approximate area of international waters in the Bering Sea

important in view of the special biological and commercial value of the area. Political change, mainly during the 1990s, resulted in a lapse in control over fisheries and in the proliferation of unregulated and illegal practices in the industry. Attempts to assess stocks, set quotas, regulate fishing and record catches are thwarted by this situation. Estimates of the scale of illegal activity vary widely, but all have one thing in common, which is their magnitude. These estimates not only represent the chaotic nature of fisheries in the region, but lost revenue for the Government of the Russian Federation, further increasing the challenge of managing marine resources.

INTRODUCTION

It is no longer remarkable to state that certain fish stocks in a given region are in decline: half of the world's major fishery resources are now in urgent need of remedial management (Anon., 1999). The Bering Sea is one such region, but also one where precious diversity remains, a fact that elevates it to the status of a priority region for conservation at a global level. This much has been recognized by several prominent conservation organizations, including WWF and The Nature Conservancy (TNC).

WWF has identified 200 priority ecoregions globally which comprise the most outstanding terrestrial, freshwater and marine habitats, known collectively as the *Global 200*. Selected for their importance in terms of biodiversity, it is this criterion that delimits these regions, not necessarily geographical boundaries. The Bering Sea ecoregion is one of 61 marine ecoregions designated by WWF. The aim of an ecoregional approach to conservation is to sustain the maximum range of biodiversity through international co-operation. It is a basic premise of ecoregional conservation that any shift in the status of one part of the system - for example in the fish on which birds and mammals depend - has potential to affect the whole.

More information is required to understand better the conservation priorities for the Bering Sea ecoregion and information has been lacking for the Russian side of the area, in particular. As part of a strategy to help conserve the Bering Sea, TRAFFIC carried out investigative work into industrial fishing in the Russian zone of the Bering Sea. The work was aimed at a better understanding of the impact of exploitation of its marine resources, for the purpose of developing recommendations and strategies to redress problems of over-harvest and illegal practice in the fisheries sector. References to catches from the north-east Pacific area, which includes the eastern Bering Sea, are made in the spirit of ecoregional conservation, as a means of keeping in mind that pressure on stocks is not limited within any geographical or political boundary. This study aimed to explore the:

- historical and economic characteristics of the fisheries in the Russian Far East;
- relevant administrative, legislative and enforcement controls on fisheries;
- key fisheries of the area, including trends in production and trade;
- illegal fishing practices in the region and factors conducive to these.

It is hoped the findings presented here will contribute to the overall body of knowledge about the Bering Sea and, most importantly, assist fisheries managers to achieve the goal of ensuring that the use of Bering Sea marine resources is at sustainable levels.

METHODOLOGY

This project was co-ordinated by the TRAFFIC Europe - Russia Office and research was conducted by TRAFFIC staff and consultants, including one based in Kamchatka, (see **Figure 1**). Research was concentrated over a six-month period commencing in July 1999, although desk-based research continued during 2000 and 2001. It entailed review of available literature and collection of pertinent trade data from selected government agencies, as well as field investigations.

Literature Review: Published and "grey" literature was reviewed, as were available local and international fishery statistics from soucres such as the Kamchatka Regional Directorate for Protection and Reproduction of Fish Stocks and Regulation of Fishing (*Kamchatrybvod*), the Food and Agriculture Organization of the United Nations (FAO) and the National Oceanic and Atmospheric Administration (NOAA) of the USA. The Kamchatka-based researcher also had access to classified information.

Field Research: Numerous interviews were conducted at sea and on land in the Kamchatka region during the latter half of 1999. Interviews were conducted by Russian researchers. No undercover tactics were used in this investigation, but owing to the professional position of one researcher, access to personnel concerned with fisheries issues was particularly easy.

At sea: A two-week trip was made on board an inspection vessel, the *Ingeneer Martynov*, patrolling in the Russian part of the Bering Sea in late July/early August. The area surveyed was in the southern Bering Sea, to the east of the Kamchatka Peninsula. Eight fishing and fish-processing vessels were inspected by the patrol during the trip. Participating in the inspection team's activities, one of the researchers for this report was able to interview captains of the vessels inspected, production managers, representatives of companies and sailors, amounting to eleven people in total. The crew and both inspectors aboard the *Ingeneer Martynov* were also interviewed. All the crew members had previously worked on fishing vessels. They and other interviewees were asked for information on the status of fisheries and illegal fishing practices in the Bering Sea.

On shore: Interviews were carried out with officials from relevant agencies such as *Kamchatrybvod*; the State Customs Committee; the Regional Directorate of the Federal Security Service; the Special Sea Inspection of the Regional Environmental Committee and the local media. These interviews were aimed at collecting Customs data and gathering information on fisheries, trade in fisheries products and management techniques, such as quotas. Additional interviews specifically targeted fishing companies and appropriate individuals located in selected harbours.

Credit: Alexey Vaisman, TRAFFIC Europe-Russia

View of the Kamchatka shore from the *Ingeneer Martynov*. Note the dented and rusted side of the vessel: enforcement equipment for policing fisheries in the Russian Far East is generally considered to be inadequate for the task.

During the survey as a whole, information was sought on evidence of violations, the types of violations that may be encountered and the methods used to conduct illegal activities. It is important to note that there were often serious constraints on collecting such information and on examining the Russian fishing industry in general, owing to the sensitive nature of the subject. Several contacts asked that their names be kept confidential. In order to protect these sources, information in this report is frequently cited anonymously as "TRAFFIC survey data". Information from other people who requested that they not be mentioned personally has been referenced by institution rather than by individual.

It has been necessary to limit the focus of this study and, as such, the issue of exploitation of salmon species in the Bering Sea has not been addressed in this report, nor have coastal and river poaching been touched upon. It should not be inferred, however, that these issues are of lesser importance for conservation in the region.

Definitions

Russian EEZ (Exclusive Economic Zone) - that part of the sea which is within 200 nautical miles from the Russian coast, yet beyond the territorial waters which extend 12 nautical miles from shore.

Russian Far East - For the purposes of this report, this is defined as those regions of the Russian Federation forming its eastern seaboard (Primorsky Kray; Khabarovsk Kray; Kamchatka; Magadan; Sakhalin; Koryak and Chukotka). Three of these border the Bering Sea.

Credit: Alexey Vaisman, TRAFFIC Europe-Russia

Petropavlovsk-Kamchatskii, the fishing centre of Kamchatka.

Kamchatka - Kamchatka is the representative province of the Russian Far East for the purposes of this report, as it is where research was concentrated. After Primorsky Kray, it is the next-most important area in terms of pollack fishing in the western Bering Sea: Primorsky Kray received 63% of the pollack quota for the western Bering Sea in 1996, Kamchatka 22% (Pautzke, 1997).

It is important to note that Kamchatka is bordered by both the Bering Sea and the Sea of Okhotsk. Therefore, when statistics for Kamchatka are referred to, it should be remembered that these comprise catches not only from the Russian part of the Bering Sea, but also from the other waters of the Russian Far East, notably the Sea of Okhotsk.

HISTORY OF INDUSTRIAL FISHERIES IN THE RUSSIAN PART OF THE BERING SEA

Industrial fishing in the Bering Sea can be traced back over a hundred years (Cline and Williams, 2000). At the end of the nineteenth century, the Kamchatka Trade Society, a private commercial company with government support, was established with the aim of trading, on both domestic and foreign markets (Japan and America). During this same period, the Japanese fishing industry also began to focus intensively on the fish resources in the waters of Kamchatka, including those of the Bering Sea, to the east of the peninsula. Such activity increased with the signing in 1907 of a Russian-Japanese fishing convention, in accordance with which the Russian Government authorized Japanese fishermen to catch and process fish and other sea products (except for seals), along the coasts of far east Russia. New species not previously harvested, or only lightly harvested, were introduced into commercial exploitation and while catches of cetaceans and pinnipeds had been the most important for fishermen at the turn of the century, by the end of the 1930s catches of fish and invertebrates had taken precedence.

The first Soviet assessments of fish stocks in the western Bering Sea were not conducted until the **1930s**, at which time they did not, apparently, reveal substantial stocks worthy of commercial exploitation by trawlers (Anon., 1996a). Hence the first Soviet trawlers constructed in the late 1920s and early 1930s (Haskell, 1963: in Pautzke, 1997), focused fishing effort in the eastern Bering Sea. By the **1950s**, Soviet fisheries policy evolved into an era of development and expansion, exemplified by increased investment in vessel construction and research on harvest technologies and processing (Anon., 1996a). Industrial fishing intensified in the early 1950s in coastal Soviet waters and in the late 1950s extended to the high seas (Zilanov, 1996: in Pautzke, 1997). By **1963** the Soviet Union reportedly had the largest and most modern fishing fleet in the world (OES, 1963 in Pautzke, 1997). In the western Bering Sea, crab, herring and Pacific Ocean Perch *Sebastes alutus* stocks had experienced significant declines by the mid-1960s (Anon., 1996a). In the **1970s**, pressure on already declining fish stocks escalated following the expansion of national economic zones at sea. The Soviet Union first claimed its EEZ in the late 1970s, while declaration of the US EEZ in 1977 excluded Soviet (and other) fishermen from that area. In response to there being fewer accessible fishing grounds around the world, Soviet trawlers returned to their own waters and fishing effort in the western Bering Sea reportedly tripled (Anon., 1996a). According to Bizikov (1996), fish stocks depleted in the 1950s and 1960s had been recuperating slowly at least until this point, but had not yet reached their former levels.

Soviet fisheries activities in the **1980s** can be described as relatively stable, made more so by State-allocated financial support in the form of grants, capital investments, subsidies and loans. In Kamchatka, this development was manifested in the emergence of a strong "single-industry" (fishing), which consisted of large fish processing companies and fishing co-operatives with coastal fishing fleets and processing plants. The late 1980s also witnessed the emergence of joint ventures, set up between Russian enterprises and foreign partners. These joint ventures aimed at producing and exporting products that would earn hard currency. The majority of foreign partners were companies from the East Asian region - Japan, China and the Republic of Korea (South Korea) -, but partnerships also developed with American, German, Norwegian and Icelandic companies and others. The first joint venture in Kamchatka was established in **1989** (Anon., 1999b).

Overview of the fishery in the 1990s

The Soviet State support apparent in the 1980s largely ceased with the political changes of the early 1990s. Fishing enterprises found themselves not only cut off from State support, but also burdened by the huge expenditures incurred by keeping old, economically unprofitable vessels. Higher prices for fuel and other resources and materials increased running costs in the fishing industry and this brought about a reduction in the number of fishing vessels in operation (Anon., 1999b). As a result, the Kamchatka fisheries sector, like the rest of the Russian economy, has been in crisis throughout the 1990s. The first stage of structural reform (1992-96) led to a significant decrease in production potential and a worsening of the industry's financial situation. Catch and production fell by more than 25% in 1992 in the Kamchatka region and continued to fall so that by 1994 output by the fishing companies in the region had halved (Anon., 1999b).

Those fishing companies remaining in business switched their focus to produce goods, often new products, for which there was foreign demand, in order to generate hard currency.

The Government of the Russian Federation regards the second stage of reforms during the 1990s (1997-2000) as having been a transition to economic growth through changes in production that were both passive (curtailing production of goods for which there is no market demand) and active (bolstering production capacity for competitive and mainly new products). The volume of hard currency-earning exports registered by the Kamchatka Customs Administration rose several-fold between 1997 and 1999 (TRAFFIC survey data). For example, the production of fillets, minced Alaska Pollack *Theragra chalcogramma*, crab claws, salmon roe and liver, commodities in demand in Asian markets, rose four- to five-fold in the mid-1990s (e.g. see **Table 11**). In contrast, production of frozen unprocessed fish fell to half the volumes of the Soviet era (Anon., 1999b). Export-oriented operators within the industry geared up to this change by investing in modern mid-sized trawlers equipped with filleting lines and roe-collecting machines, as well as crab packaging and salmon processors. Some State support was available to assist regeneration of the national fishing industry in the form of monetary policies, for example a reduction in the rate of VAT (value added tax) from 16-20% to 10%. Administrative measures put in place to assist the industry included the redistribution of quotas in favour of small entrepreneurs.

Joint ventures, in particular, were set up for the production and export of hard currency-earning products. Following the establishment of the first joint venture in Kamchatka in 1989, other such partnerships were formed, with interests in more than 25 countries, especially in the Asia-Pacific region. In 1993, there were 14 joint ventures in Kamchatka harvesting fisheries resources and another 14 involved in their processing. However, there was a distinct swing towards joint ventures in export trade in the mid-1990s and by 1995 only two ventures continued to harvest fish and seafood, while 11 were still in business processing catches, but the number of joint ventures involved in trade had risen from 28 in 1993, to 37 in 1995 (TRAFFIC survey data). Despite private ownership, many of the joint ventures remained economically dependent on the former State structures and support in practice (Anon., 1999b).

One effect of the shift to production for export was an increase in trans-shipment at sea. The advantage of this for fishing companies is that a return to port may be avoided and with it the need to pay VAT and transport and other expenses. In 1995 and the years thereafter, the joint ventures and all major exporters in the region, continued to increase the volumes that they traded outside the country's twelve-mile territorial water limit, thereby circumventing the costs associated with landing catches in the Russian Federation. In 1995, the volume of the catch from beyond the 12-mile limit was four times as high as the volume that cleared Customs. The price per kilogramme of an exportable catch in 1999 was reported to be one-third of what it was in 1995. This may have been in part owing to the export of lower-value, raw products instead of processed commodities, following the destruction of local canneries. The cheaper way of exporting from outside the territorial waters may also account for lower unit prices for exports (see **Types of violations**), but these prices would not be reflected in Customs records.

For businesses unable to enter foreign markets, the shift in trade was sometimes in quotas rather than fish. This trade took on truly massive proportions. According to *Kamchatrybvod*'s inspectors, "They used to sell fish; now they sell quotas. Three or four years ago the Russian Federation stuck to the position of trading fish abroad. Now Russia's fishing sector is selling its quotas to foreign States as a direct industry." (Sviridova, 1998).

Among the casualties of the transition of focus to export were the Kamchatka-based canneries. They had been set up for mass production for domestic consumption and relied on hiring seasonal workers during the peak salmon runs. In the early 1990s, the production of canned seafood collapsed over a very short time as suppliers of raw and semi-processed fish to the Kamchatka coastal-based fisheries industries became market-led (Anon., 1999b). The production of canned seafood in Kamchatka slumped by 90% in the early 1990s and the system of supply to coastal processing plants was destroyed.

Many of the changes in the fishing industry in the 1990s were "survival mechanisms", means of keeping qualified personnel employed. However, although there was a rise in catches and quotas rose in the middle of the decade - for example, the fish catch rose in 1995 and 1996 (Anon., 1999b) - and although more than 70% of the total production was officially exported in 1996, Kamchatka's fishing economy ended that year in the red. Developed within the framework of a single-industry economy based on fisheries, Kamchatka could not react quickly enough to the transformations in the market that accompanied the dismantling of the Soviet Union. Despite investments by the federal government, the Bering Sea fisheries sector generated only about 150 billion roubles (RUR) in 1996 (USD1 equivalent to approximately RUR4000-4500 during that year) (Anon., 1999b) and the fisheries industry of Kamchatka, which had long been the region's mainstay, accounted for only 9.5% of Kamchatka's budget revenue in 1998. According to figures for that year, 71% of Kamchatka fisheries enterprises and canneries were losing profits and the entire industry continued to lose money, despite a few growth areas (Anon., 1999b). Matters worsened after the Russian Federation was struck by a financial crisis in 1998, defaulting on its loans to foreign creditors. Purchase tax on frozen fish products rose from 10% to 20% on 1 July 1999 (Anon., 1999c). In the light of these adverse conditions, the Russian Government has reportedly recognized the State has a renewed role to play in regulating economics and in the fishing sector. Measures for State control over market processes in the fisheries sector include re-allocation of catch quotas and devices to increase the efficiency of natural resource use (Anon., 1999b). It is against this background that the industrial fisheries of the Russian sector of the Bering Sea have been operating during the 1990s and within this context that the following sections of this report should be considered.

STRUCTURE OF FISHERIES ADMINISTRATION

Fisheries in the Russian Federation are administered at federal level by the State Committee of Fisheries, which is an entity with the same status as a ministry within the government of other nations. The aim of the Committee is to enhance, preserve and use rationally the living marine resources of the Russian Federation and to develop the fishing industry. In order to achieve its objectives, the Committee is assisted by its regional branches, or *rybvods* (in Kamchatka this is *Kamchatrybvod*), and scientific/research institutions, for example, the Russian Federal Research Institute of Fisheries and Oceanography (VNIRO). The Committee also consorts with other federal agencies (Anon., 1999d). The functions of the State Committee of Fisheries include:

- formulation, implementation and monitoring of fisheries policies and regulations;
- analysis, co-ordination and oversight of the socio-economic, scientific and technical aspects of the fishing industry;
- short- and long-term development of the fishing industry;
- co-ordination of scientific research pertaining to stocks, stock enhancement, and harvest levels;
- supervision of fishing ports;
- monitoring safety on marine vessels;
- granting licences to fishing fleets and fish hatcheries;
- confirmation of final catch quotas;
- monitoring use of catch quotas;
- determination of opening and closing dates of fishing seasons, zones and grounds.

FISHERIES LEGISLATION AND AGREEMENTS

National legislation

Fisheries activities in the Russian Federation are controlled legislatively by *The Fisheries Act, 1989*. This legislation is under review and subject to revisions at present. Additionally, fisheries are regulated by various decrees issued by the President or Government (signed by the Prime Minister). A Decreee of the Government of the Russian Federation especially relevant to this report is that which governs export of fisheries product direct to a foreign port from the Russian EEZ (beyond territorial waters). As such trade is not subject to the same Customs regulations which apply within Russian territory, the decree requires captains of vessels wishing to transport marine products harvested from the Russian EEZ direct to a foreign port to complete a Customs declaration before the export is made.

Some fisheries matters are under the power of the State Committee of Fisheries and its regional Directorates. A decree of this Committee, for example, prohibited the catching of Alaska Pollack from 1 March to 20 April 2000, during its spawning season (Anon., 2000p).

International agreements

A number of current international agreements have direct and indirect bearing on the fisheries sector and fishing activities in the western Bering Sea and these include the following.

- ***Convention on the Conservation and Management of Pollack Resources in the Central Bering Sea*** This Convention applies to the high seas area (referred to as the Convention Area, or more commonly, the Donut Hole) of the Bering Sea, beyond the EEZs of the USA and the Russian Federation. It aims to establish an international regime for the conservation, management and optimum use of pollack resources in the Area; to restore and maintain pollack stocks at levels which will permit maximum sustainable yield; to co-operate in information gathering and examination; and to provide a forum in which to consider the establishment of necessary conservation and management measures for living marine resources other than pollack in the Convention Area, as may be required in the future. This Convention was signed on 16 June 1994 by the USA, Russia, China and South Korea, on 4 August 1994 by Japan, and on 25 August 1994 by Poland. The Convention had entered into force for all signatories by 4 January 1996 (Anon., 2000a).

  An extensive pollack fishery existed in the Donut Hole in the 1980s. The Donut Hole was heavily trawled by China, South Korea, Poland, Japan and the Soviet Union, to the point where overall annual catches plummeted from 1.4-1.5 million t in 1989 to 300 000 t in 1991 and reached a mere 11 000 t in 1992. The severity of the situation led the governments to agree to a two-year voluntary suspension of fishing (1993 and 1994). Following the signing of the Convention, the Parties agreed to maintain the fishing ban until the biomass of the Aleutian Basin pollack stock exceeded a threshhold of 1.67 million t. At a 1999 meeting of the Parties (8-12 November 1999), the scientific data available revealed that the Aleutian Basin stock was approximately 654 228 t, or 1.02 million t less than the threshold required to allow a commercial pollack fishery in the Donut Hole (Anon., 2000a). While it was clear that pollack stocks had not recovered, several Parties (Japan, China, South Korea and Poland) supported the reopening of the fishery in 2000. However, as there was no consensus on this issue, the harvest for 2000 remained at zero (Anon., 2000a).

- ***Agreement between the Government of the United States of America and the Government of the Union of Soviet Socialist Republics on Mutual Fisheries Relations*** (Basic Instrument for the US-Russia Intergovernmental Consultative Committee - ICC): This bilateral agreement, which was signed on 31 May 1988 and entered into force on 28 October 1988, has been extended twice and will remain in force until at least 2003. The Russian Federation succeeded the Soviet Union as a party to the Agreement. The Agreement commits the parties to work co-operatively on a wide variety of fisheries issues of mutual concern. The Agreement also establishes the Intergovernmental Consultative Committee (ICC), which meets annually to review issues arising under the Agreement. The ICC has become the principal US-Russian venue for considering matters of fishery conservation and management, scientific research on fisheries and co-operation on fisheries enforcement, particularly in the Bering Sea and North Pacific Ocean (David Balton, Director, Office of Marine Conservation, US Department of State, *in litt.*, 27 March 2001; Anon., 2000a).

- ***UN General Assembly resolution "Large-scale pelagic drift-net fishing, and its impact on the living marine resources of the world's oceans and seas" (A/RES/46/215)***:

  On 20 December 1991, at their 79th plenary meeting, the United Nations General Assembly established this resolution in response to deep concern about reports of expanding large-scale pelagic drift-net fishing on the high seas and a review of the best scientific information confirming the negative impact of this type of fishing on the marine environment. Accordingly, the UN General Assembly called upon the members of the international community to take the following actions:

  - achieve, by 30 June, 1992, a 50% reduction in fishing effort in existing large-scale pelagic high seas drift-net fisheries by reducing the number of vessels involved, the length of nets and the area of operation.
  - fully implement a global moratorium on all large-scale pelagic drift-net fishing by 31 December, 1992.
  - take measures independently and as a collective international community in co-operation with intergovernmental and non-governmental organizations and well-established scientific institutions to prevent large-scale pelagic drift-net fishing and to report violations to the Secretary General.

  Annually, the General Assembly has considered reports to the Secretary General from States, inter-governmental and non-governmental organizations and well-established scientific institutions concerning activities or conduct inconsistent with the terms of the moratorium, again calling on the Secretary General to monitor its implementation.

- ***United Nations Convention on the Law of the Sea (UNCLOS)*** opened for signature in 1982 and has been ratified by the Russian Federation. It entered into force on 16 November 1994 and "embodies and enshrines the notion that all problems of ocean space are closely interrelated and need to be addressed as a whole". It is globally recognized as the regime dealing with all matters relating to the law of the sea. According to UNCLOS, Coastal (member) States exercise sovereignty over their territorial sea, which they have the right to establish up to a limit of 12 nautical miles from shore, and have sovereign rights in the EEZ, which extends up to 200 nautical miles from shore. They exercise jurisdiction over marine science research and environmental protection in the EEZ and are obliged to adopt, or co-operate with other States in adopting, measures to manage and conserve living resources in the high seas. States bordering enclosed or semi-enclosed seas are expected to co-operate in managing living resources, environmental and research policies and activities (Anon., 2001).

- The ***Agreement for the implementation of the provisions of the United Nations Convention of the Law of the Sea of 10 December 1982 relating to the conservation and management of straddling fish stocks and highly migratory fish stocks*** was adopted in 1995 to set new standards for managing fish stocks in a sustainable manner. The Agreement has been ratified by the Russian Federation, but has not yet entered into force. The Agreement is premised on the need for States to co-operate in order to conserve straddling and highly migratory fish stocks. Among other provisions, the Agreement stipulates that member States shall "apply the precautionary approach widely to conservation, management and exploitation of straddling fish stocks and highly migratory fish stocks in order to protect the living marine resources and preserve the marine environment" (Anon., 2001).

The following two agreements, although they relate principally to salmonids, which are outside the main focus of this report, are included to illustrate further the nature of existing agreements between the Russian Federation and other nations for the conservation of Bering Sea resources.

- ***Convention for the Conservation of Anadromous Stocks in the North Pacific Ocean*** (Basic Instrument for the North Pacific Anadromous Fish Commission - NPAFC): Canada, Japan, the Russian Federation and the USA are parties to this treaty, which entered into force in 1992. The Convention Area for this treaty covers the high seas portions of the North Pacific Ocean and adjacent seas north of 33°N The Convention prohibits directed fishing for anadromous species (those which swim up a river from the sea to spawn) in the Convention Area and requires that the by-catch of anadromous species in other fisheries be minimized. The NPAFC has become the primary forum for the exchange of information and co-operation in dealing with illegal and unauthorized fishing in the North Pacific region. In 1999, this co-operation resulted in the detection of ten unauthorized drift-net vessels, of which three were seized (Anon., 2000a). The People's Republic of China and South Korea participate as observers at meetings of the NPAFC. One or both of them may become parties to the Convention in the future (David Balton, Director, Office of Marine Conservation, US Department of State, *in litt.*, 27 March 2001.)

- ***US-Russia Bilateral Agreement on Harvesting Salmonids within their Exclusive Economic Zones***

  In September of 1992, the USA and Russian Federation concluded this agreement, imposing a ban on directed fishing for Pacific salmons in the US and Russian EEZs, including the Bering Sea beyond 25 nautical miles of the US and Russian coasts, between 170° E and approximately 143° West, north of 50° North. With this agreement, a joint scientific programme on anadromous stocks was also established, primarily to facilitate the exchange of information on salmon. In order to co-ordinate efforts and monitor programme implementation, the agreement proposed exchanging scientific personnel between governments and conducting annual meetings in the USA and Russian Federation, alternately.

MANAGEMENT OF THE FISHERY

Stocks and quotas

Stock assessments in the Russian part of the Bering Sea are conducted by scientists from the regional scientific/research institutes of the State Fisheries Committee. In the Russian Far East these institutes include: the Pacific Research Institute of Fisheries and Oceanography (TINRO); the Sakhalin Scientific Institute of Fishing Industry (SakhNIRO); and the Kamchatka Scientific Institute of Fishing Industry (KamchatNIRO). The responsibilities of these institutes, as well as the Moscow-based VNIRO, are the following:

- assessment and monitoring of fisheries resources;
- development of allowable harvest levels; and
- development of means to restore and improve aquatic habitats (Anon., 1999d).

In addition, through various international agreements, scientists collaborate on an international level by exchanging information on current research; this collaboration assists Russian scientists in assessing and monitoring stocks and in ascertaining appropriate harvest levels (Anon., 2000a).

Comprehensive information on stocks is not available to the general public because it is regarded as confidential and/or commercially sensitive (V.I. Radchenko, TINRO, *in litt.* to TRAFFIC Europe, 2000). As mentioned in **Methods**, the stock information presented in this report was collected from various sources.

Following stock assessment by the regional research institutes, recommendations for the Total Allowable Catch (TAC) are submitted at the request of the State Fisheries Committee to VNIRO in Moscow, for compilation of the Proposed Total Allowable Catch (PTAC) for approval. Following approval, the PTAC is sent for State Environmental Review under the Ministry of Natural Resources. If agreed by the review commission, a general quota - TAC- is established, which will then be submitted for division among the regions. Any further changes in the TAC should also be subjected to an environmental review (A. Shestakov, pers. comm., 2000). Quotas set for certain species for the Russian part of the Bering Sea for the years 1996-99 are presented in **Table 1**.

The general quota is divided into the following four categories (Safronov, *Kamchatrybvod*, pers. comm. to TRAFFIC Europe).

1. An **industrial quota**, which is distributed free to domestic companies only (or those believed to be so). The recipients of this kind of quota include commercial companies (as a rule former large State companies) and local administrative bodies (TRAFFIC survey data). In 1999, Kamchatka Region Administration received allocation of an extra 1700 t for crab, supposedly because of the fuel crisis which restricted vessels' ability to venture onto the high seas (Anon., 2000b).

2. A **scientific quota**, allocated to government research institutes, which is divided into two categories:

   i) a quota for research into the current ecological, biological and reproductive characteristics and distribution of harvested species.

   ii) a "controlled catch" that is officially for testing fishing gears, for monitoring the daily movements of shoals and the "daily catch situation". In reality the "controlled catch" can be a means of generating profit to support research institutes. TINRO (the Pacific Research Institute of Fisheries and Oceanography), for example, received a crab quota of 16 000 t with a market value of USD15 million in 1996, portions of which it sold off to commercial companies for cash (Anon., 1998b). For the purposes of controlled catches, a hired vessel can usually fish anywhere, including closed districts.

   In 2000, procedures have been revised, such that research institutes are no longer in full control of their quotas, partial control having been vested in the State Unitary Enterprise "National Fish Resources" (Anon., 2000c; Safronov, *Kamchatrybvod*, pers. comm.).

3. A **quota for payment**, that is sold to domestic companies by auction.

4. A **quota under international agreements**, which can be distributed to foreign vessels in return for payment, or as compensation, for example for halting fishing activity in international waters in the Bering Sea (the Donut Hole) and the Okhotsk Sea (the so-called Peanut Hole).

Table 1
Catch quotas for the Russian part of the Bering Sea for selected species (t)

SPECIES	1996	1997	1998	1999
Alaska Pollack	476 150	573 440	508 830	656 660
Cod	61 430	74 450	45 950	43 455
Herring	26 000	23 000	58 000	236 100
Plaices (including Yellowfin Sole)	6200	7700	9920	9248
Halibuts and flounders	4850	2650	3855	3368
Saffron Cod	2900	4500	10 300	7300
Crab	2400	2180	645	640
Rockfishes	200	800	675	125
Pandalus shrimp	1200	1500	300	1000
Whelk	300	2300	1250	0

Source: *Kamchatrybvod*, 1999.

From **Tables 1** and **2**, it appears that the portions of quotas allocated to foreign fleets were small in 1995, for most resource categories, compared to those allocated to domestic users. However, South Korea's quota for 30 t of Alaska Pollack in 1995 (see **Table 2**) is dwarfed by its quota for 110 000 t of the species from the Russian Bering Sea in 2001, as reportedly announced by South Korea's Ministry of Maritime Affairs and Fisheries. This amount is itself a vast increase on the corresponding quota of 35 600 t for 2000. It is reported that South Korea bid USD108/t to secure the 2001 quota (Anon., 2001). In June 2001, it was reported that the Russian Federation had raised USD160 million through the sale of quotas in a round of auctions (Anon., 2001).

Table 2

Fishing quotas awarded to foreign fleets in the Russian Bering Sea, 1995, (t)

	Alaska Pollack	Cod	Halibut	Squid	Crabs	Other
Japan	3800					
South Korea	30	800	30			30
Japan	1100					
South Korea	30	800	30			3
USA			500			
USA			500			60
Japan	2900					
Japan	400					
Japan	10 000					
Japan	680	850	340		70	340
Japan				2150		2000
TOTAL	**18 940**	**2450**	**1400**	**2150**	**70**	**2433**

Source: *Kamchatrybvod*, 1999.

Table 3

Records of foreign fishing vessels in the Russian Bering Sea, 1995

Species fished	Months of the year	Type of vessel	Nationality	Tonnage	Number of vessels
Salmon	5-7	Driftnetters	Japanese	127	53
Salmon	5-7	Driftnetters	Japanese	127	7
Cod	5-12	Longliners	Japanese	250	7
Cod	8-10	Longliners	South Korean	350	2
Alaska Pollack	6-12	Trawlers	Japanese	270	1
Alaska Pollack	7-12	Trawlers	Japanese	300	1
Alaska Pollack	5-10	Factory ships	South Korean	2700	1
Halibut	8-9	Longliners	US	100	3

Source: *Kamchatrybvod*, 1999.

During this survey, it was found that quotas are often re-sold, although this practice is prohibited by Russian law (see **Violations**). Quotas can also be altered. In March 2000, for example, it was reported that the overall quota was to be changed by allocating 3000 t of king crab quota to a company (apparently a Moscow-based enterprise without boats or ships), to the detriment of pre-existing quota allocations in the Russian Far East (Anon., 2000d).

Finally, it should be noted that Russian fishermen also receive quotas to fish in the waters of other nations and in 1998 such quotas amounted to roughly five or six million tonnes. Recently, only about 25-30% of these allocations have been used, because of a lack of suitable ocean-going vessels and funds to finance longer fishing trips. In 1998, the take from foreign quotas was expected to be about 1.3 million tonnes (Anon., 1998c).

Stakes are high in the struggle for quotas for the most valuable seafood species within the Russian Federation and human lives may even be risked. In an incident reported in the summer of 1999 in the newspaper *Rossijskaya Gazeta*, directors of three large fishing companies died as a result of violent events related to the division of quotas.

ENFORCEMENT OF FISHERIES LEGISLATION

Enforcement of fisheries regulations is carried out by a wide variety of agencies. These include *Kamchatrybvod*, the Special Marine Inspection (SMIS) of the former State Committee for Environmental Protection (now part of the Ministry of Natural Resources), and the Division of Sustainable Use and Protection of Marine Biological Resources of the Regional Directorate (RIOMBR) of the Federal Border Service. Other agencies involved in enforcement are the Customs Service, the Federal Security Service, the Ministry of Interior and the Federal Service of Tax Police. These bodies employ conventional methods to monitor fishing activities, i.e. they inspect vessels using patrol boats, and conduct aerial patrols of marine fishing areas. In 1997, the Sea Patrol division of the Federal Border Service had about 1000 patrol boats, ships and auxiliary vessels for use in anti-poaching operations (Safronov, *in litt.*, 1999; Mitin, 1997).

One of the main ways to monitor the activities of foreign fishing vessels in the Russian EEZ is the permanent presence of a fisheries inspector (observer). These inspectors check that the catch is in accordance with the allocated quotas. They also verify that foreign fishermen are complying with conditions stipulated in the fishing licence and with the provisions of *The Fisheries Act, 1989*, and subsidiary legislation regulating fishing in Russian waters. This daily (**continuous**) monitoring is applied to foreign vessels only, whereas all other monitoring applies to vessels of any nationality. This, and other monitoring measures, are described below.

Continuous monitoring

Fisheries inspectors are permanently based on each foreign fishing vessel as an observer, but not on Russian vessels. They may either be fisheries enforcement agency staff or freelance observers. Their duties include:

- the daily monitoring of catch per species to check for compliance with quotas assigned to the vessel;
- the daily monitoring of compliance of foreign fishermen with the fishing regulations in force within the Russian EEZ;
- the daily collection of all information on fishing activity;
- the transfer of the information gathered during what is referred to as the "inspection hour", to a designated senior inspector.

The senior inspector oversees a group of vessels for which a daily inspection at a specified time is organized. Communication is by radio on a frequency specially assigned to this purpose. The senior inspector gathers operational information from observers on all vessels in his group and then collates the information for reporting to the Regional Fisheries Inspectorate.

Periodic monitoring

This is carried out by State fisheries inspectors on patrol ships, who board fishing vessels to inspect them. These visits may be scheduled or unscheduled. If weather conditions permit, monitoring involves comprehensive examination of a vessel's compliance with fisheries regulations. Periodic monitoring also entails examination of the work of the permanent observer based on the vessel under inspection. Specifically, an inspection would consist of checking:

- documentation relating to the vessel's fishing activities;
- holds and containers (through a careful comparison of their measurements with the original construction plans of the ship);
- the fishing gear, to ascertain compliance with requirements stipulated in the fishing licence and in any relevant inter-governmental treaty;
- compliance of the vessel with all requirements stipulated in current fisheries legislation;
- the composition of the catch to ascertain compliance in terms of species and size;
- the percentage of the allowable catch that has been recorded in the vessel's documentation.

Credit: Alexey Vaisman, TRAFFIC Europe-Russia

A fisheries inspector prepares to be hoisted on board a processing ship to make his checks.

Targeted monitoring

Targeted monitoring is carried out at specified checkpoints by government inspectors from various agencies. These inspectors check vessels that are *en route* to a harbour to unload processed fish products.

Aerial patrolling

Aerial patrolling of marine areas within the Russian EEZ is one of the most important and effective ways of monitoring fishing activities. It can provide evidence of illegal fishing in prohibited waters and deter fishing vessels from entering such areas. *Kamchatrybvod* uses its own helicopters for aerial patrols or leases AN-24 aircraft fitted with extra fuel tanks.

Exceptional control operations

Exceptional control operations have been organized and undertaken by a multi-agency team since 1994. These are called *putina*, from the Russian for "fishing season". These annual expeditions are major enforcement operations staged by the *rybvod*, Special Marine Inspection, and the Federal Border Service. Their aim is to crack down on Russian and foreign operations that may be catching or exporting fish from the Russian EEZ illegally. All fishing vessels and processors in a particular area will have inspectors present during a *putina* (Pautzke, 1997).

RECORDED CATCHES IN THE RUSSIAN PART OF THE BERING SEA

Figures for catch volumes, according to various categories, from the Russian part of the Bering Sea are presented in **Tables 4**, **5** and **6**.

Catch data are not readily available for recent years. One reason for this is that catches are reported to regional *rybvods*, not by area of catch, but area of registration of the fishing company in question. That means that catches harvested under the quota for the Russian part of the Bering Sea could be reported in any one of several *rybvods* - for example, in the ports of Magadan, Khabarovsk, Vladivostok, or Petropavlovsk-Kamchatsky - depending on the where the vesssel's company was registered (TRAFFIC survey data). This makes for a complex system, where harvests cannot begin to be matched against quotas before a process of collation of records from the various administrations in the region. Moreover, vessels reporting catches in these *rybvods* do not necessarily fish only in the western Bering Sea, but operate also in other fishing grounds, particularly the Sea of Okhotsk. The result is a protracted means of collecting and separating out catch statistics relating specifically to the Russian Bering Sea.

One of the Major Fishing Areas defined by FAO for statistical purposes, the Pacific Northwest (Area No. 61), encompasses the western Bering Sea within its vast area, which stretches south beyond the Tropic of Cancer (see **Figure 2**). Catch data for the Pacific Northwest are presented, although it is a much larger expanse of ocean than the subject of this report, to provide additional perspective on the area's fishery for selected species (see **Annex 2**).

Table 4

Recorded Russian catches (t) of fish and other seafood from the Russian part of the Bering Sea, compared with catches for the Russian Far East and for the whole of the Russian Federation, 1990-99

Year	Overall Russian catch	Russian Far East	Western Bering Sea
1990	10 389 000	4 601 000	1 184 440
1991		4 060 000	873 710
1992		3 159 400	912 700
1993	4 368 700	2 778 300	936 580
1994	3 542 500	2 318 700	
1995	4 236 900	2 804 100	
1996			*581 630
1997	4 537 500	3 146 400	*692 520
1998	4 353 500	3 018 300	*639 725
1999	4 100 000		*957 896

Source: "Ministry of Fisheries" (Moscow), as cited in Anon., 1996a; Zilanov, 1996; Monakhov, 1998; *Rybnoye Khozaistvo*, in Anon., 1999e and in Anon., 2000e.

* Catch data are unavailable for recent years, therefore the figures presented for 1996 to 1999 show minimum quota volumes for the Russian part of the Bering Sea, as calculated from **Table 1**.

Figure 2

Map to show FAO fishing area no. 61 - the Pacific Northwest (ocean shaded light grey). Note that area no. 61 includes, but extends far beyond, the western Bering Sea

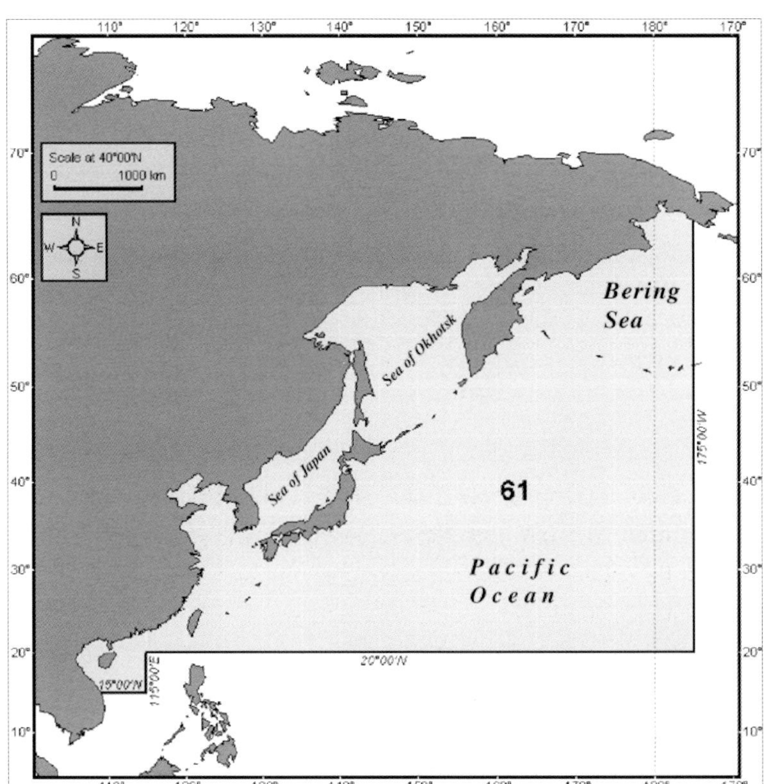

Source: Food and Agriculture Organization of the United Nations.
Note: Map based on Miller cylindrical projection

Table 5

Recorded catches (t) by foreign fleets in 1997

Fishing area	Alaska Pollack	Cod	Halibut	Squid	Plaice	Rockfish	Other	Total
Russian part of the Bering Sea	24 271	7735	995	1939	612	231	603	36 386
Off coast of S.E. Kamchatka	2137	113	10	15	37	0.6	52	2365

Source: *Kamchatrybvod*, 1999.

Table 6

Recorded catches (t) in the western Bering Sea, 1980 to 1993

Year	Pollack	Herring	*Pleuro-nectidae	POP	Saffron Cod	Salmon	Cod	Halibut	Prawn	Crab	Other	Total
1980	928 008	12 800	20 814	1200	14 021	14 770	14 181	301	300	800	40 835	1 048 030
1981	890 943	14 906	10 612	1797	13 430	55 508	33 102	6589	530	1970	71 253	1 100 640
1982	1 019 120	12 880	12 010	540	12 870	18 800	62 160	2940	-	2990	68 510	1 212 820
1983	970 950	16 260	17 210	100	15 390	47 640	63 640	2010	-	3530	66 690	1 203 720
1984	785 890	17 440	8000	60	16 120	29 360	97 460	2640	-	3220	26 960	987 150
1985	712 800	31 310	33 460	40	10 270	38 400	94 870	2860	-	3170	18 100	945 280
1986	936 690	20 980	39 900	20	8890	24 120	117 650	5030	90	4800	15 120	1 173 290
1987	1 108 300	20 200	24 000	200	9600	52 400	72 400	4400	3500	3200	29 600	1 327 800
1988	1 291 700	15 340	27 900	980	10 480	21 750	70 340	2520	-	5810	238 570	1 685 390
1989	1 213 800	9470	24 010	40	9770	65 470	61 950	2820	10	4500	113 090	1 504 930
1990	928 400	16 270	26 760	50	15 220	16 500	89 180	2980	-	4290	84 790	1 184 440
1991	631 460	12 180	29 000	30	7490	96 140	61 820	1530	-	3860	30 200	873 710
1992	702 710	2370	25 450	30	13 520	29 880	110 000	1380	-	0.6	26 790	912 700
1993	768 840	2040	11 410	1100	5170	59 130	62 080	0.3	-	1410	25 080	936 580

Source: "Ministry of Fisheries" (Moscow), in Anon., 1996a.

Note:* Pleuronectidae in this table = flounders and plaices other than halibuts, e.g. Yellowfin Sole *Limanda aspera*; POP = Pacific Ocean Perch, a rockfish *Sebastes* spp.; Saffron Cod = Far Eastern Navaga *Eleginus navaga*

FISHING GEARS USED IN THE RUSSIAN PART OF THE BERING SEA

Drift-nets

Drift-nets are widely recognized as having a very negative impact on the environment (see UN resolution in **International agreements**). Despite this, the use of drift-nets is permitted for research-oriented "controlled catch" (see **Stocks and quotas**) (*Kamchatrybvod* staff, *in litt.*, 18 June 2000) and also for Japanese vessels with quotas for salmon, for the purpose of catching such fish only. The mesh size of salmon drift-nets ranges from 50 to 70 mm, their height is five metres and the length of one unit is 50 m. Tens to hundreds of units are joined together and may form nets several kilometres long that often snare a high percentage of by-catch. There is no limit for the length of a drift-net, only for the distance between drift-nets; these limits vary depending on the fishing area and the species targeted (*Kamchatrybvod* staff, *in litt.*, 18 June 2000). The Russian Federation and the USA have agreed to prohibit the use of drift-nets in international waters in the Bering Sea, i.e. the Donut Hole.

Bottom-nets

Flatfish that live on the bottom of the sea, such as halibuts and flounders, are the main target of bottom-nets. Russian specialists report that huge damage is caused by this type of gear because of its lack of selectivity (high percentage of by-catch) and because of the destruction of benthic biotopes (the sea bottom and its associated organisms) when the net is freed from its moorings in rough weather conditions. The net becomes entangled in seagrass and seaweed, causing vast destruction of this vegetation and high mortality of invertebrates and young fish that live in this rich and protective environment. No legislation regulates the use of bottom-nets in the Russian Federation (*Kamchatrybvod* staff, *in litt.*, 18 June 2000).

Bottom seine or *snuurevaad*

Bottom seines are widely used for catching pollack and all so-called bottom-living fish species. In theory, a bottom seine has to be led close to the sea bottom, but cannot touch it. In practice, seines are purposely taken right to the sea bottom to pick up crabs. Bottom seines are a common form of gear for poaching and it has been reported that they can be used as a kind of pelagic trawl and are occasionally used to catch a shoal of salmon, instead of the official target fish for such gear (e.g. pollack or flounder) (TRAFFIC survey data).

Table 7

Selected types of fishing gears and the species for which they are permitted in the Russian Bering Sea

Gears	Fish
Drift-net	salmon
Bottom-net	cod, halibut, flounders and plaices
Bottom seine (*snuurevaad*)	cod, pollack, plaices
Bottom trawl	Illegal in Russian waters
Pelagic trawl	pollack, rockfishes, (shrimp?)
Long-lines	rockfishes, halibuts, cod
Traps and pots	crabs
Seine	herring
Pound net	herring
Kiddle	salmon

Source: Kamchatrybvod; extracts from regulations, orders and internal documents.

THE FISHERY SPECIES-BY-SPECIES

The following section describes the most important commercial marine resources targeted by industrial fisheries in the Russian part of the Bering Sea, as identified through literature searches and consultations with specialists. Information on the biology, stocks, fishery, trade and trends of each type of fish, as it relates to the region, is provided. The Russian name for each fish or other species is included in brackets after the English common name in the sub-headings of this chapter.

Alaska Pollack *(Mintay)*

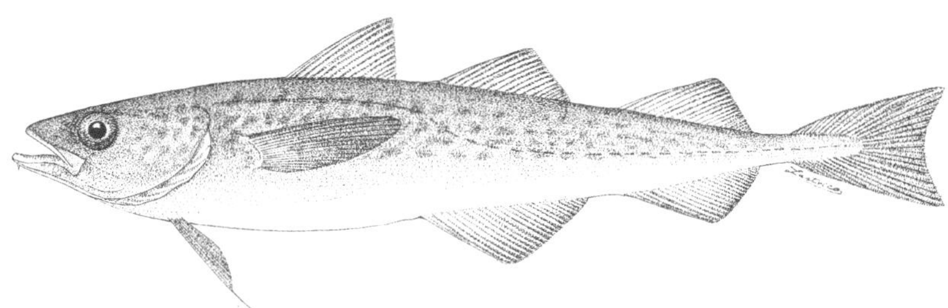

Alaska Pollack *Theragra chalcogramma*

Credit: Food and Agriculture Organization of the United Nations, Rome, Species Identification and Data Programme, Fisheries Department.

Alaska Pollack *Theragra chalcogramma* can live to an age of 15 years and reach a maximum length of 91 cm and maximum weight of 1.4 kg. Reproductive age is at about three years, by which time a body length of 22 to 28 cm has been reached and by six years of age, pollack are usually 40 to 48 cm long. In the western Bering Sea, the fish begin to spawn from the end of March to early April. The main spawning region is Olyutorsky Bay. A large part of the spawn develops at the surface and drifts with the currents to the feeding shallows of the Karaginsky Bay (Anon., 2000f; KamchatNIRO staff, pers. comm., 2000).

The Alaska Pollack is the dominant fish species among those targeted by industrial fisheries in the Bering Sea. It accounts for about half the biomass of groundfish in the Sea and is the basis for one of the largest single-species commercial fisheries in the world (Cline and Williams, 2000). This makes the pollack a significant species, not only from an economic point of view, but also as an important factor for social stability in the region (KamchatNIRO staff, pers. comm., 2000). The importance of Alaska Pollack to the industry has been the cause of major disputes between fishing fleets of various countries.

Stocks

In the western Bering Sea, separate stocks of Alaska Pollack are found in Olyutorsky Bay and further north, along the Navarin shelf (Ianelli and Wespestad, 1998). This latter ("northern") stock reproduces in the eastern areas of the sea and moves into the Russian EEZ before and during winter hibernation. It spreads into western waters up to 176°E. The stock in the vicinity of Olyutorsky Bay ("western stock") ranges as far as 180°E. Larger stocks of the species are found in the eastern Bering Sea and in the Sea of Okhotsk (Ianelli and Wespestad, 1998; Zaitsev, 1996).

The western stock of pollack gradually grew from the early 1970s until 1982, from one to 3.2 million tonnes, following which peak the stock steadily declined (Balykin, 1986; 1990). Diminishment of the stock was caused mainly by two factors: poor stock recruitment (insufficient young joining the breeding populations) and unregulated fishing in international waters of the Bering Sea. This latter factor was especially significant following adoption of US legislation giving US fishing activities priority in the US EEZ, to reduce competition from non-US fishing and processing concerns in the eastern Bering Sea. As a result, fishing vessels from countries without a Bering Sea coastline shifted focus from the US EEZ towards the Donut Hole. It should be recognized, however, that Alaska Pollack is a species that exhibits considerable fluctuation in its stock size, owing to natural population dynamics. Some ichthyologists believe that the stock is experiencing a natural decline at present, as well as the effects of fishing pressure.

According to the estimates of TINRO specialists, the total stock of the western Bering Sea pollack aged from two to nine years during the period 1991-96 was as shown in **Table 8**. From the table, it can be seen that the decrease in biomass shows a marked and continous decline. Even an increase in numbers of shoals in 1994 did not halt this decline, indicating a rise in the proportion of young fish. The minimum size of Alaska Pollack for permitted harvest in Russian waters of the Bering Sea is 32 cm.

Table 8
Alaska Pollack stock (aged two to nine years) in the western Bering Sea

	1991	**1992**	**1993**	**1994**	**1995**	**1996**
Numbers (billions)	6.73	6.08	4.52	6.10	4.48	
Biomass (million tonnes)	1.8	1.6	1.4	1.3	1.12	1.02

Source: Pacific Research Institute of Fisheries and Oceanography (TINRO), unpublished.

Fishery

Industrial exploitation of Alaska Pollack began at the end of the 1960s. In the 1970s, the species constituted 85 to 90 percent of the total catch of fisheries products taken in the western Bering Sea (Anon., 1996a). In 1977, the USA introduced its 200 nautical mile EEZ, causing a sharp increase in pollack fishing in the Bering Sea outside this limit. Consumer demand for Alaska Pollack, particularly more recently for *surimi* and fish roe, the rapid development of industrial fishing techniques, the increase in number of fishing vessels and their specialization in pollack fishing have all contributed to continued high catch levels for the species in the western Bering Sea. Catches of Alaska Pollack from the western Bering Sea, 1980-93 are shown in **Table 6**.

For the Northwest Pacific Ocean, catches of Alaska Pollack reported to FAO have averaged around four million tonnes annually for the period 1984-98 (see **Annex 2**), with a decline noted in the late 1990s from 3 450 800 t in 1995, to 2 266 200 t in 1999. The Russian Federation records the largest catches in the Pacific Northwest, and Russian catches show declines in line with overall declines for the whole region (see **Annex 2**). The catch of Alaska Pollack from the Pacific Northeast in 1999 was only 1 096 250 t, according to statistics collected by FAO.

Bottom seine is the gear most usually used for catching Alaska Pollack, but mid-water pelagic trawls are also employed. As mentioned above (**National legislation**), a ban on pollack fishing was imposed for the period 1 March to 20 April 2000 in the Karaginsky subzone of the western Bering Sea, since this is a period of active spawning (Anon., 2000g; Anon., 2000h).

Alaska Pollack fishing is plagued by illegal activity (see **Violations**). The actual volume of Alaska Pollack that is harvested, according to *Kamchatrybvod* staff, is 150% of the quota. Estimates from the Interior Ministry and the Federal Security Service for Kamchatka Region differ, pitching the unaccounted overrun of the quota at 15% (*Kamchatrybvod* staff, pers. comm., 1999).

Trade

Alaska Pollack was the leading fisheries product exported from Kamchatka in terms of volume in the second half of the 1990s. The cumulative volume exported from Kamchatka between the start of 1995 and the second quarter of 1999 was 157 500 t (see **Table 9/Figure 3**). Exports have risen from around 2500 t in 1995, to 46 500 t in 1997, to 78 500 t in 1998 (see **Table 9**). Countries importing Alaska Pollack from the Kamchatka region have grown from two in 1995 (China and Ukraine), to a total of 15 for the period 1995-98 (see **Table 9**).

The development of consumer demand for new products, such as *surimi*, fillet and roe, has acted as a powerful stimulant for the recent development of the pollack fisheries, whereas previously the cheaper form of headed fish was the main form marketed. The Russian Federation and the USA are the main suppliers of Alaska Pollack roe to the Japanese market and, together with Poland, these countries supplied 54% (49 885 t) of the Alaska Pollack roe appearing on the Japanese market in 1997 (*Kamchatrybvod*, unpublished). However, the market for these products is under pressure, largely because the resource has declined. Russian exports of this type of product to Japan fell by 30% in 1996, by comparison with 1995 exports (see **Table 10**). This was despite the fact that the *surimi* production capacity rose, with an increase in the number of factory ships in the latter half of the 1990s.

Table 9

Alaska Pollack *(Mintay)* exports (all forms of fish) from the Kamchatka region, 1995-March 1999

	1995		1996		1997		1998		Jan.-March 1999		Av. annual export '95-98		
	t	USD x1000	t	USD x1000	t	USD x1000	t	USD x1000	t	USD x1000	t	USD x1000	USD/kg
China	954.8	369.6	1244.1	543.7	19 512.9	9777.9	15 603.6	7856.8	6339.4	2653.6	9328.8	4637.0	0.50
Cyprus	0.0	0.0	0.0	0.0	0.0	0.0	3087.7	1958.7	1742.0	1343.1	771.9	489.7	0.63
France	0.0	0.0	0.0	0.0	0.0	0.0	884.0	1142.3	0.0	0.0	221.0	285.6	1.29
Germany	0.0	0.0	688.9	915.0	6474.5	5059.3	4342.2	6364.6	11 505.5	12 338.9	2876.4	3084.7	1.07
Hong Kong	0.0	0.0	0.0	0.0	280.1	131.7	4474.0	3050.2	70.1	10.9	1188.5	795.5	0.67
Japan	0.0	0.0	0.0	0.0	325.0	108.5	5627.0	4087.8	3.4	0.7	1488.0	1049.1	0.70
S. Korea	0.0	0.0	426.5	168.6	2878.3	1088.8	14179.0	8338.3	4552.1	2647.9	4370.9	2398.9	0.55
Liechtenstein	0.0	0.0	0.0	0.0	1093.0	855.8	0.0	0.0	0.0	0.0	273.2	214.0	0.78
Panama	0.0	0.0	0.0	0.0	0.0	0.0	258.5	169.3	0.0	0.0	64.6	42.3	0.66
Poland	0.0	0.0	0.0	0.0	400.0	218.0	0.0	0.0	0.0	0.0	100.0	54.5	0.55
Singapore	0.0	0.0	0.0	0.0	15 584.3	8293.0	6607.0	4264.7	5901.9	3930.9	5547.8	3139.4	0.57
Thailand	0.0	0.0	0.0	0.0	0.0	0.0	1000.0	572.0	0.0	0.0	250.0	143.0	0.57
Ukraine	1583.1	571.8	0.0	0.0	0.0	0.0	1737.8	436.0	3.5	1.4	830.2	252.0	0.30
USA	0.0	0.0	0.0	0.0	8.6	3.8	10 384.5	7819.1	2713.8	5052.7	2598.3	1955.7	0.75
Virgin Is.	0.0	0.0	0.0	0.0	0.0	0.0	10 337.6	6277.6	6194.8	4063.6	2584.4	1569.4	0.61
TOTAL	**2537.9**	**941.4**	**2359.5**	**1627.3**	**46 556.6**	**25 536.7**	**78 522.8**	**52 337.4**	**27 521.0**	**19 704.7**	**32 494.2**	**20 110.7**	**0.62**

Source: Kamchatka Customs, 1999.

Figure 3
Alaska Pollack exports from the Kamchatka region

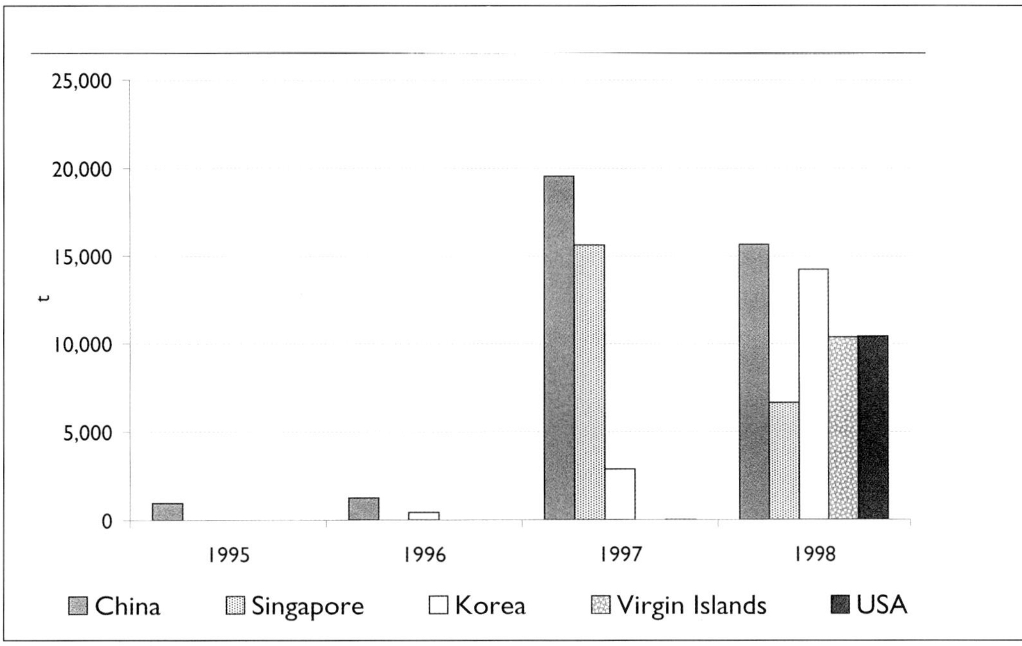

Source: Kamchatka Customs, 1999. **Note**: Korea = South Korea.

Table 10
Deliveries of Alaska Pollack mince from the Russian Federation to Japan, 1992-97

Year	Volume (t)	Value (JPY10^6)	Mean price, JPY/kg
1992	26 961	11 770	436.6
1993	16 702	3922	234.8
1994	17 306	3762	217.4
1995	22 071	5486	248.6
1996	15 715	3819	243.0
1997	12 649	4339	343.0

Source: *Kamchatrybvod* staff, pers. comm., 1999.

Production of Alaska Pollack products newly in demand which use only part of the whole fish, such as roe and fillets, have resulted in increased dumping of pollack discards into the sea, by Russian, American, Japanese, Chinese, Polish, and South Korean vessels sailing in the western Bering Sea (TRAFFIC survey data).

Trends

The outlook for pollack fishing in Russian part of the Bering Sea appears unfavourable in the coming years. Despite the fact that stocks of pollack in the Russian part of the Bering Sea have been in decline since 1982 and that catches have been declining, the quota for the Russian part of the Bering Sea has risen steadily from 1996 to 1999 (see **Table 1**). The level of harvest, both legal and illegal is estimated to be too high, stimulated by strong demand, especially for the more processed forms of the fish on Asian markets.

What is *surimi*? *Surimi*, or "washed fish mixture" translated from Japanese, has been produced and eaten in Japan for over 1000 years. Traditionally, *surimi* consists of fresh fish, starch products, herbs and flavourings. It is currently used to manufacture imitation crabmeat, shrimp, and other marine products (Schulte-Paul, 1999). Made by a process of "de-watering", *surimi* quality depends on the freshness of the fish and the rapidity of the processing. Fillets are ground into paste and washed repeatedly; when only protein solids remain, sugar and gelling agents are added (Anon., 1999f; Hodgson, 1992). *Surimi* processing on vessels started in the 1960s and expanded after 1975, largely due to the discovery of new freezing technology that allowed freezing of fish (minus the head, guts and backbone) on board without quality loss. This new technique allowed fish to be frozen in the fishing region and then processed further elsewhere without sacrificing quality. The main species used for *surimi* is Alaska Pollack, although since 1990 technical developments have allowed use of additional species (Anon., 1999g; Schulte-Paul, 1999). Nevertheless, Alaska Pollack is considered to produce the best-quality *surimi* (Anon., 2000i).

Crabs

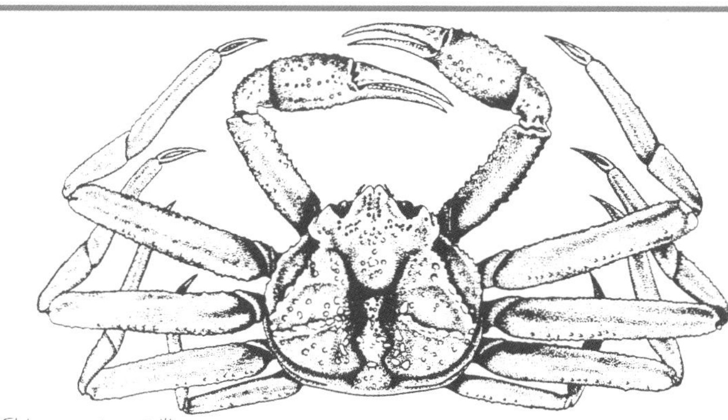

Snow Crab *Chionoecetes opilio*

Credit: Food and Agriculture Organization of the United Nations, Rome, Species Identification and Data Programme, Fisheries Department.

In the Bering Sea live several species of crab that are significant commercially: the **Kamchatka** or **Red King Crab** ***Paralithodes camtschatica***; **Blue King Crab** ***P. platypus***; **Stony King Crab** (also known as Golden King Crab, Even-spine Crab or Brown Crab) ***Lithodes aesquispinus***; **Scarlet King Crab** ***L. couesi***; **Snow Crab** ***Chionoecetes opilio*** and **Tanner Crab** ***C. bairdi***. Species of crab that occur in the cold waters of the Bering Sea are in general long-lived, slow-growing and late-maturing. Mass migration is an important biological characteristic of the crab. Movement from one part of the sea to another is induced by seasonal changes and crabs move constantly within the limits of their distribution in search of optimum environmental conditions, including food. Such movements can cause a patchy, inconsistent distribution of crabs (Safronov, *in litt.*,1999).

Crabs are highly fecund. Female Kamchatka King Crabs, for example, can produce 81 000-200 000 eggs (Anon., 1994b). Once hatched, the natural mortality rate of young crabs can reach 96% and more (Safronov, *in litt.*,1999).

Kamchatka King Crab: This species commands the highest price of all the crabs in the western Bering Sea. The average shell width for this species is 16 cm, although specimens have been known to reach 25 cm. Legs often reach one metre in length but can grow to one-and-a-half metres. Adult Kamchatka King Crabs generally weigh between two and four-and-a-half kilogrammes each (Anon., 1995), although specimens of over 10 kg that might be 20 to 30 years of age have been recorded (Blau, 1997).

In the western Bering Sea, Kamchatka King Crabs are found in the highest concentrations in the Olyutorsky-Karaginsky area of the Kamchatka shelf (Zilanov *et al.*, 1989, in Anon., 1996a) but the greatest quantity in the Bering Sea as a whole is found in Bristol Bay (in the eastern part of the Sea). Kamchatka King Crabs can be found at a range of depths between four and 250 metres (Safronov, *in litt.*,1999).

Credit: Caroline Raymakers, TRAFFIC Europe

Kamchatka King Crab and Gold King Crab meat and frozen salmon roe on display at the Seafood Expo, Brussels, Belgium, in 2001.

Blue King Crab *Paralithodes platypus*: This species is the most prevalent among those harvested in the western Bering Sea. Multiple populations of Blue King Crabs are concentrated in the north-west parts of the Bering Sea (Safronov, *in litt.*,1999).

Stony King Crab *Lithodes aequispinus*: The Stony King Crab is a deep-water species and is relatively new to the market. Exploitation began in the 1980s, as an incidental catch to the Kamchatka King Crab (Anon., 1996a).

Scarlet King Crab *Lithodes couesi*: This species is found on muddy bottoms at depths of 140 to 1000 m. Fecundity varies with the size of the crab, but a female can produce 4000 to 5000 eggs (McCrae, 1994).

Snow Crab *Chionoecetes opilio* and **Tanner Crab *Chionoecetes bairdi***: The common name of "tanner crab" can refer to these two species. In Olyutorsky Bay, an increase in fishing intensity has had extreme negative effects on the populations living there (Safronov, *in litt.*,1999).

Stocks

The main resources of crabs in the Bering Sea are in US waters. Only Blue King Crabs occur in higher numbers in the western part of the Bering Sea, where they are also more widespread (Shuntov *et al.*, 1995). Stocks of Blue King Crab in the region of the Koryak shelf have gradually reduced in recent years. According to specialists, this is the result of intensive catching and environmental changes. In the Navarin region, the condition of the population of Blue King Crabs is currently stable and adequate, because of an absence of fishing (Safronov, *in litt.*,1999).

In recent years, populations of Snow Crabs and Tanner Crabs have been relatively stable, although their numbers are very small. Consequently, it is essential to regulate the catch of these species thoroughly (Safronov, *in litt.*,1999).

The abundance of Stony King Crabs is not known (Anon., 1985).

Fishery

In the western Bering Sea, catches of crab have declined almost continuously, from 5810 t in 1988, to 1410 t in 1993 (see **Table 6**). Recent catch quotas for the western Bering Sea (645 t in 1998 and 640 t in 1999) are much lower than those of 1996-97, which were in excess of 2000 t.

Catch statistics reported to FAO for the whole of the Pacific Northwest indicate that the Russian Federation caught no king crabs in the area between 1984 and 1999 (see **Annex 2**).

King crabs are usually caught in baited pots.

Trade

Crabs are the most expensive fisheries products traded from Kamchatka (see **Table 21**). Kamchatka's fishermen have a saying, which is, "We didn't till, didn't sow, but caught crabs - so there's money in our pockets." Since the early 1990s, Japan has been the major importer of

Russian crab, with much smaller quantities going to France and the USA (see **Table 11**; Anon., 1999g). Reported exports of crab from Kamchatka are as shown in **Table 12** for the period 1995-99. About 15 000 t out of a total of 16 770 t of crab exported by Kamchatka was reportedly destined for Japan.

In 1994, Russian deliveries of crabs for export exceeded the 10 000 t mark, probably partly explicable by a ban on king crab fishing in Alaskan waters in 1994. In 1995, almost the entire harvest went for so-called "special-purpose imports", meaning to specific buyers, and there were almost no crabs on the open market (TRAFFIC survey data).

According to certain importers, Russian Kamchatka King Crab prices hit a low in 1995, after which they began to rise in line with reduced quotas for the species in the western part of the Bering Sea (see **Crabs**, ***Fishery***).

Table 11
Reported imports (1000 t) of crab into Japan, USA and France from the Russian Federation, 1990-98

Product	1990	1991	1992	1993	1994	1995	1996	1997	1998
Fresh king crab to Japan	0.1	0.3	1.5	2.2	4.7	8.2	16.1	20.4	19.1
Frozen king crab to Japan	3.0	3.0	5.1	8.9	18.8	24.2	29.3	22.6	20.5
Fresh Snow Crab to Japan	0.0	0.3	1.4	2.9	6.7	9.9	13.0	15.7	16.5
Frozen Snow Crab to Japan	4.9	6.1	5.7	3.6	4.2	12.8	11.3	11.8	6.2
Canned crab to France	.791	.791	.716	.791	.631	.336	.180	.066	.029
Frozen crab to USA			2.0	1.4	2.7	3.9	7.4	11.8	13.6
TOTAL	**8.791**	**10.491**	**16.416**	**19.791**	**37.731**	**59.336**	**77.28**	**82.366**	**75.929**

Source: Anon., 1999e.

Table 12
King crab exports (all forms) from the Kamchatka region

	1995		1996		1997		1998		1999 Jan.-Mar.		Average annual export '95-98		
	t	USD x1000	t	USD x1000	t	USD x1000	t	USD x1000	t	USD x1000	t	USD x1000	USD /kg
Canada	0.0	0.0	0.0	0.0	42.9	334.1	0.0	0.0	0.0	0.0	10.7	83.5	7.80
Japan	5132.3	62438.2	2591.5	22554.7	4503.4	32175.6	2808.0	19076.1	491.3	3436.3	3758.8	34061.2	9.06
S. Korea	0.0	0.0	240.6	2083.1	137.6	1015.8	889.7	8606.7	66.0	464.3	317.0	2926.4	9.23
USA	0.0	0.0	126.7	1076.8	104.5	672.6	193.3	1099.0	31.7	259.0	106.1	712.1	6.71
Total	**5132.3**	**62438.2**	**2958.8**	**25714.6**	**4788.3**	**34198.1**	**3891.0**	**28781.9**	**589.0**	**4159.6**	**4192.6**	**37783.2**	**9.01**

Source: Kamchatka Customs, 1999.

Figure 4
King crab exports (all forms) from the Kamchatka region

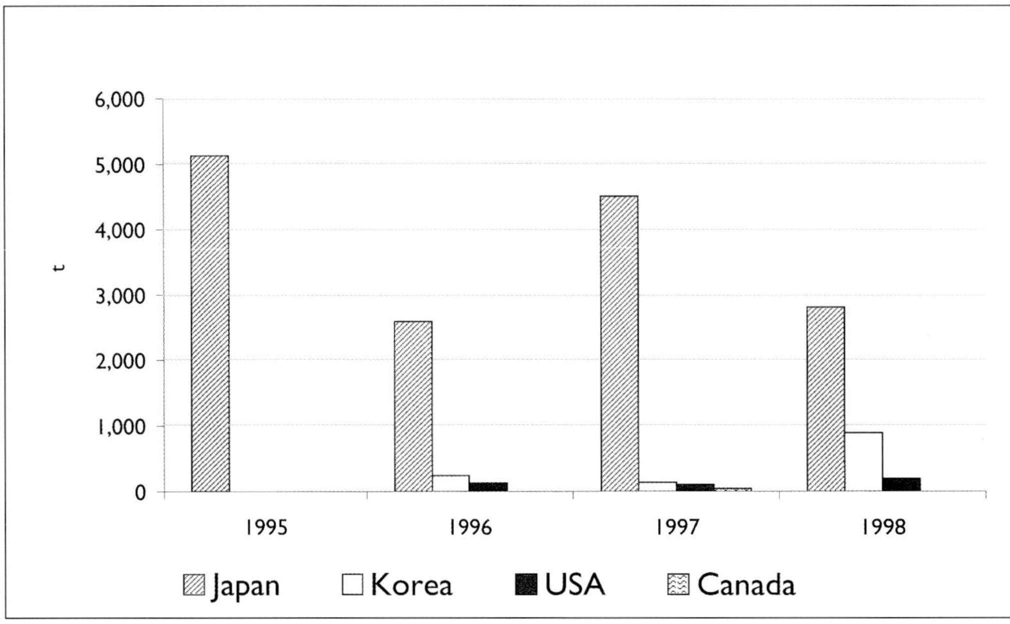

Source: Kamchatka Customs, 1999. **Note**: Korea = South Korea.

Trends

Trends for crab species are unclear because of a lack of specific information. Catches show declines in the most recent years for which data is available for the western Bering Sea and much reduced quotas for the last two years of the 1990s suggest evidence of reduced stocks. Exports from the Russian Federation to France, Japan and the USA have increased steadily over eight years in the 1990s (see **Table 11**), but for Kamchatka, no clear export trend is apparent (see **Table 12**).

Pacific Cod, Saffron Cod (*Treska, Navaga*)

There are two species of cod recorded as exported from Kamchatka - **Pacific Cod** *Gadus macrocephalus* and **Saffron Cod** *Eleginus gracilis* (also known as Far Eastern Navaga).

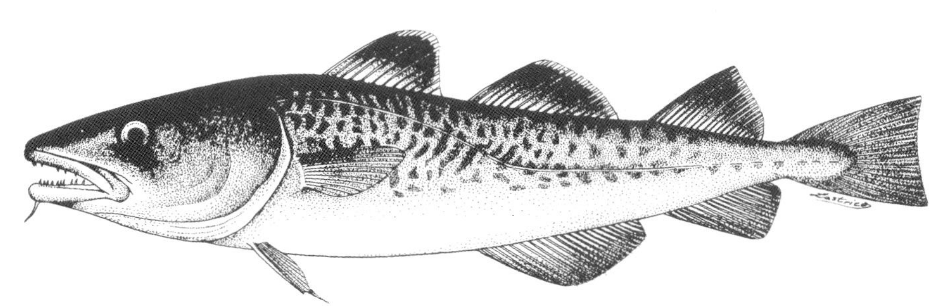

Pacific Cod *Gadus macrocephalus*

Credit: Food and Agriculture Organization of the United Nations, Rome, Species Identification and Data

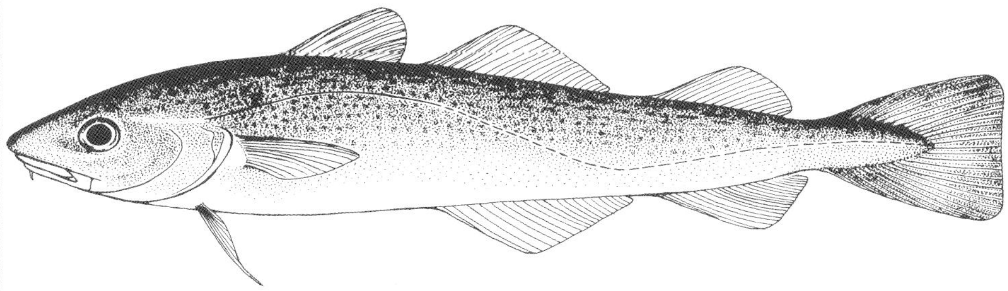

Saffron Cod *Eleginus gracilis*

Credit: Food and Agriculture Organization of the United Nations, Rome, Species Identification and Data

Pacific Cod are indiscriminate predators and occur primarily on the continental shelf and upper slopes of the Sea. They can grow to a maximum length of 117 cm, weigh up to 23 kg and live for 13 years (Anon., 2000j).

Saffron Cod occur in shallow coastal waters and may enter rivers. They are opportunistic feeders. The biggest individuals reported have been up to 55 cm in length and just over one kilogramme. Fish of 12 years of age have been recorded (Anon., 2000k), but most of the fish caught are between two and six years of age, measuring 30-70 cm and between half and four-and-a-half kilogrammes.

Mass sexual maturity occurs at five to six years of age in both species. Fecundity is related to size, and numbers of eggs per fish can vary between 0.7 million and about seven million. The spawning season extends from winter to early spring. Spawning in the Russian Bering Sea is reported to begin in March or April (at 150-370 m depth) and migration to feeding grounds on the shallow shelves takes place in May to June (KamchatNIRO staff, pers. comm., 1999).

From November to April, wintering cod accumulate in depths between 150 and 410 m and most high catches recorded have been at depths of between 180-350 m.

Stocks

In the Bering Sea, Pacific Cod is most abundant in the east. Within the western part of the Sea, the species is most abundant in the northern Navarin area. According to Ianelli and Wespestad (1998), Pacific Cod biomass in the western Bering Sea has fluctuated at around 100 000 t for the period 1978-98. The total Saffron Cod stock in this part of the Sea is thought to amount to about 85 000 t, with 50 000 t of this quantity occurring in Kamchatka province (Zaitsev, 1996).

Fishery

The industrialization of the cod fisheries in the Anadyr-Navarin region started in 1968 with the introduction of large-scale trawl fishing. At the start of the 1970s the trawler was replaced by the bottom seine (or *snurrevaad*) for catching cod: the catch of cod by bottom seine is made from May or June to October. Usually the whole year's limit is caught in this period. From 1992, long-line fishing of cod began and this is now the main method of specialized cod catching (*Kamchatrybvod*, unpublished, 1999).

For the western Bering Sea, catch data for Saffron Cod and cod (Pacific Cod are not specifically mentioned) are recorded by the State Committee of Fisheries in Moscow (see **Table 6**).

In the Pacific Northwest, three nations report catches of Pacific Cod - Japan, South Korea and the Russian Federation. The Russian Federation records diminishing catches on the whole for the period 1984-98, although the 1998 catch represents a 17% increase on the 1997 catch and, at 101 929 t, the catch for 1999 represents a further increase, by 7.5%, on the 1998 amount. South Korea, which catches by far the smallest quantity in the Pacific Northwest, exhibits increasing catches (see **Annex 2**). Catches from the Pacific Northeast, which are at much higher volumes, have been fairly steady during the 1990s (see **Annex 2**).

In 1987 (27 929 t), 1996 (21 110 t) and 1999 (47 032 t), the catch of Saffron Cod as recorded in the FAO Yearbook of Fishery Statistics was taken entirely by the USSR/Russian Federation (Anon., 2000k; see **Annex 2**).

Trade

Pacific Cod is caught and traded in larger quantities than Saffron Cod and exports of the former from Kamchatka have risen steadily from 7038 t in 1995 to 18 567 t in 1998 (**Table 13**). South Korea and China have consistently been the biggest importers. The species is marketed in fresh, frozen, dried and smoked form (Anon. 2000j).

Exports of Saffron Cod from Kamchatka are variable in quantity and destination (see **Table 13**).

Trends

Catch data for Saffron Cod in the western Bering Sea are variable and no clear trend can be identified (**Table 6**). This is further exemplified by a variable catch quota for the species (**Table 1**). The western Bering Sea catch quota for Pacific Cod declined in the last two years of the 1990s (**Table 1**): catches over the period 1980-93 fluctuated widely and no trend is discernable from these (**Table 6**). The fishing of cod is relatively insignificant compared to pollack and in various years catches have amounted to 9-15% of the latter.

Table 13

Pacific Cod (*Treska*) exports (all forms of fish) from the Kamchatka region

	1995		1996		1997		1998		Jan.-March 1999		Av. annual export '95-98		
	t	USD x1000	t	USD x1000	t	USD x1000	t	USD x1000	t	USD x1000	t	USD* x1000	USD/kg
Canada	0.0	0.0	0.0	0.0	0.0	0.0	635.2	223.2	0.0	0.0	158.8	55.8	0.35
China	466.3	30.5	1246.9	134.6	2202.8	1749.7	7041.1	3754.0	760.1	456.0	2739.3	1526.1	0.56
Japan	776.7	388.9	574.0	463.6	284.3	247.4	1309.3	1016.5	0.6	0.7	736.1	626.1	0.85
Hong Kong	0.0	0.0	0.0	0.0	47.5	7.1	8.0	5.2	114.4	28.0	13.9	3.1	0.22
Germany	0.0	0.0	0.0	0.0	0.0	0.0	47.3	37.8	0.0	0.0	11.8	9.5	0.81
S. Korea	5122.3	3454.6	7548.4	5573.6	12 727.1	7931.9	8668.7	5477.4	7072.5	4260.4	8516.6	6026.3	0.71
Norway	0.0	0.0	0.0	0.0	823.1	716.1	0.0	0.0	32.9	20.8	205.8	179.0	0.87
Portugal	583.0	378.9	0.0	0.0	0.0	0.0	0.0	0.0	0.0	0.0	145.7	145.7	0.00
Singapore	0.0	0.0	0.0	0.0	249.7	174.7	139.9	129.4	0.0	0.0	97.4	76.0	0.78
Ukraine	89.7	34.8	0.0	0.0	0.0	0.0	88.8	20.3	0.0	0.0	44.6	27.5	0.62
USA	0.0	0.0	301.3	2.0	148.2	134.7	628.2	503.5	688.4	711.5	269.4	160.0	0.59
TOTAL	**7038.0**	**4287.6**	**9670.6**	**6173.7**	**16 482.7**	**10 961.5**	**18 566.4**	**11 167.4**	**8668.7**	**5477.4**	**12 939.4**	**8835.2**	**0.68**

Table 14

Saffron Cod (*Navaga*) exports (all forms of fish) from the Kamchatka region

	1995		1996		1997		1998		Jan.-March 1999		Av. annual export '95-98		
	t	USD x1000	t	USD x1000	t	USD x1000	t	USD x1000	t	USD x1000	t	USD x1000	USD/kg
China	37.9	18.9	37.9	18.9	0.0	0.0	841.1	206.7	25.4	8.4	229.2	61.1	0.27
Japan	1878.2	1205.3	0.0	0.0	0.0	0.0	300.0	220.0	0.0	0.0	544.5	356.3	0.65
Hong Kong	0.0	0.0	0.0	0.0	0.0	0.0	51.0	10.2	1415.4	247.7	12.8	2.6	0.20
S. Korea	0.0	0.0	0.0	0.0	92.9	23.7	363.6	92.9	0.0	0.0	114.1	29.1	0.26
Singapore	0.0	0.0	0.0	0.0	68.9	24.8	719.2	290.9	0.0	0.0	197.0	78.9	0.40
Ukraine	0.0	0.0	0.0	0.0	0.0	0.0	2285.7	675.5	404.9	92.5	571.4	168.9	0.30
USA	0.0	0.0	0.0	0.0	1.4	0.7	53.4	27.7	2.0	0.5	13.7	7.1	0.52
TOTAL	**1916.0**	**1205.3**	**37.9**	**18.9**	**163.3**	**285.6**	**4614.0**	**1524.0**	**1847.7**	**349.1**	**1682.8**	**758.4**	**0.45**

Source for both tables: Kamchatka Customs, 1999

Figure 5

Pacific Cod exports (all forms of fish) from the Kamchatka region

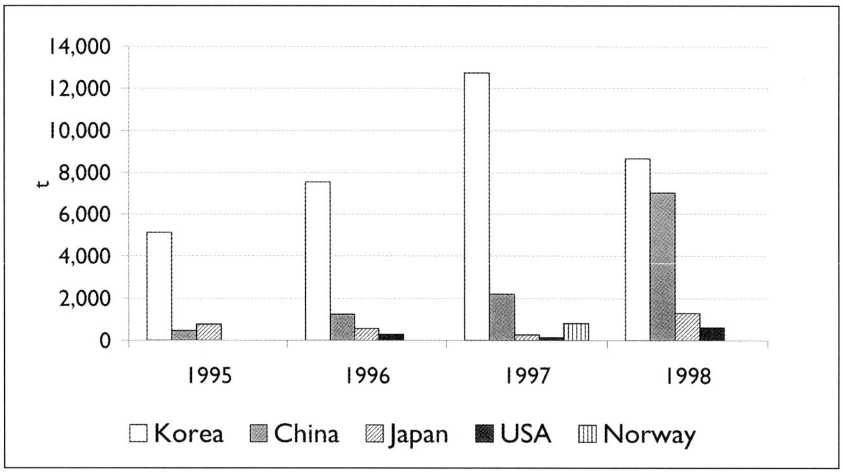

Figure 6

Saffron Cod exports (all forms of fish) from the Kamchatka region

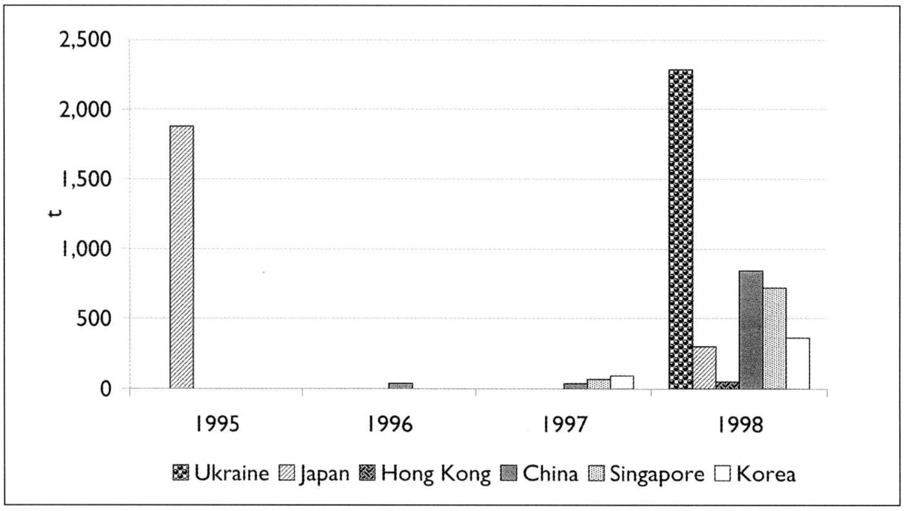

Source for both figures: Kamchatka Customs, 1999. **Note**: Korea = South Korea.

Pacific Herring *(Seld)*

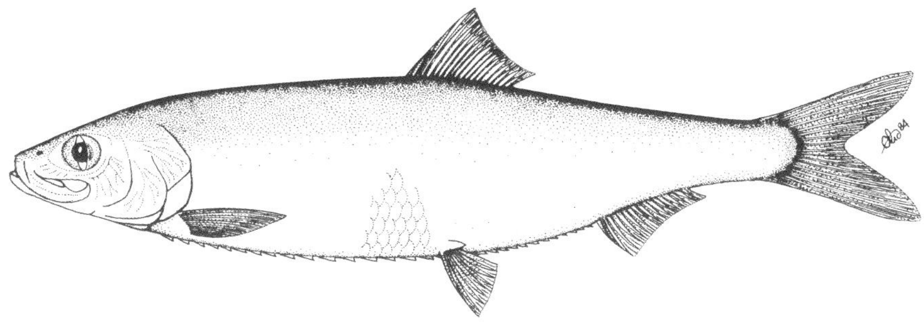

Pacific Herring *Clupea pallasii pallasii*

Credit: Food and Agriculture Organization of the United Nations, Rome, Species Identification and Data Programme, Fisheries Department.

The Pacific Herring *Clupea pallasii pallasii* is widely distributed in the waters of the Russian Far East. A gregarious species, the Pacific Herring spends most of its life here in the shelf zone, except for its second and third years, when it moves into deeper waters (Anon., 2000l). In the western Bering Sea, a local stock of Pacific Herring spawns in the Korfo-Karaginsky area. Spawning occurs in northern parts of Karaginsky Bay in the littoral zone. Spawn settles on seaweed or algae. Immediately after spawning, the herring begins intensive fattening and migrates to the waters adjacent to Olyutorsky and Karaginsky Bays before moving north-eastwards to the region off Cape Navarin. In autumn, reverse migration to Olyutorsky Bay begins and the fish spend winter here (Naumenko and Bonk, 1999a; 1999b).

Pacific Herring can reach up to 46 cm in length and live to be 19 years old. The body length of a middle-aged herring is typically between 24 and 38 cm and most mature fish weigh between 200 g and 400 g (Anon., 2000l).

Stock

The western Bering Sea stock is reported to be 150 000 t in the Korfo-Karaginsky area (Zaitsev, 1996). Naumenko and Bonk (1999a; 1999b) believe that herring stocks are increasing.

Fishery

Industrial exploitation of herring began in the second half of the 1930s and from this time until the 1950s, herring was caught with drag-nets during the spawning periods. Subsequently, drift-netting of wintering aggregations of fattening herring started, which was later replaced by fishing with purse drag-nets, and then with trawlers. At the start of the 1960s, the catch of herrings by the USSR and Japan reached 300 000 t/year. In these years, the stock of herring started to decline sharply (Kachina, 1981). The situation worsened to the point where the stock entered a state of extreme decline and in 1970 a fishing ban was introduced, which continued until 1985. These measures had positive results and scientists have continued to monitor the condition of stock and conclude that in recent years numbers of herring have increased owing to the introduction into stock of fish of middle-yield generations (Naumenko and Bonk, 1999a; 1999b).

Table 15
Herring (*Seld*) exports (all forms of fish) from the Kamchatka region

	1995		1996		1997		1998		Jan.-March 1999		Av. annual export '95-98		
	t	USD x1000	t	USD x1000	t	USD x1000	t	USD x1000	t	USD x1000	t	USD x1000	USD/kg
China	4958.8	2657.9	1456.3	780.8	3376.1	916.5	8667.6	1302.6	1298.2	619.3	4614.7	1414.4	0.31
Japan	2629.3	1081.3	5167.8	3422.1	13913.0	6776.7	8192.9	3248.7	240.2	109.5	7475.8	3632.2	0.49
S. Korea	0.0	0.0	0.0	0.0	20.9	6.3	787.8	169.3	1200.0	396.0	202.2	43.9	0.22
Ukraine	538.4	326.3	0.0	0.0	0.0	0.0	3881.9	453.1	1391.6	459.2	1105.1	194.9	0.18
USA	40.0	15.3	0.0	0.0	31.4	3.7	0.0	0.0	0.0	0.0	17.9	4.7	0.26
TOTAL	**8166.5**	**4080.7**	**6624.1**	**4202.9**	**17 341.5**	**7703.2**	**21 530.2**	**5173.7**	**4130.1**	**15 84.0**	**13 415.6**	**5290.2**	**0.39**

Figure 7
Herring exports (all forms of fish) from the Kamchatka region

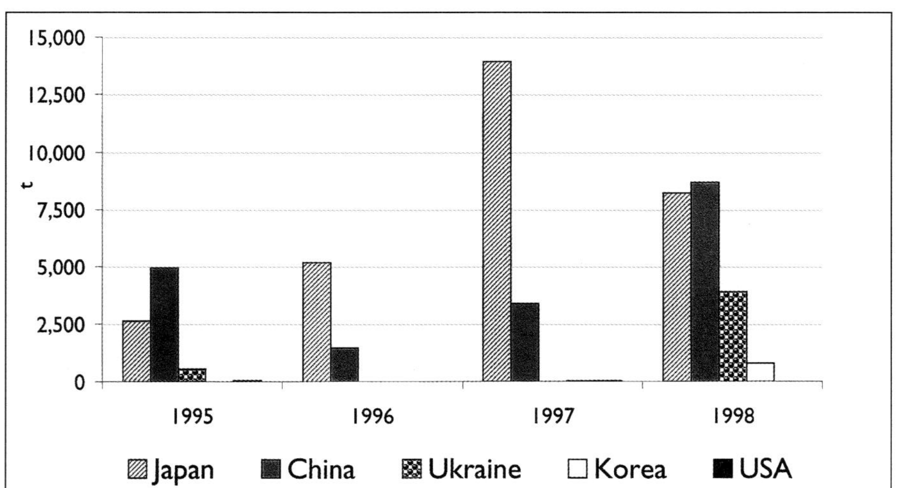

From 1990-93, catches of the species in the western Bering Sea declined (**Table 6**), despite rising quotas, but Russian catches in the Pacific Northwest overall have increased dramatically since 1994, for example by 26% from 1997 to 1998 (see **Annex 2**). According to FAO statistics, the Russian catch of herring for 1999, however, had decreased by 9% relative to the 1998 catch.

Herring in the Russian part of the Bering Sea are harvested in the present day using pelagic trawls and seines.

Trade

Exports of Pacific Herring from Kamchatka rose steadily for the period 1995-98 (see **Table 15** and **Figure 7**). Five countries are recorded as importers, with Japan importing the largest amounts during this time period.

Trends

Stocks are thought to be increasing in the western Bering Sea and it is reported that from 1994 to 1996, 25% of the total catch quota remained unallocated because the fleet did not have the necessary capacity to harvest it. A change in herring fishing may be heralded by the fact that the larger, fatter, well-fed herring caught by trawling in the winter are now less favoured than leaner, younger fish and roe, for which demand is increasing on markets in south-east Asia. Consequently, scientists recommend a review of the strategy for herring fishing that would allow a significant increase in catches in the spawning period (Naumenko and Bonk, 1999a; 1999b, Safronov, pers. comm., 2000).

Rockfishes (*Okun*)

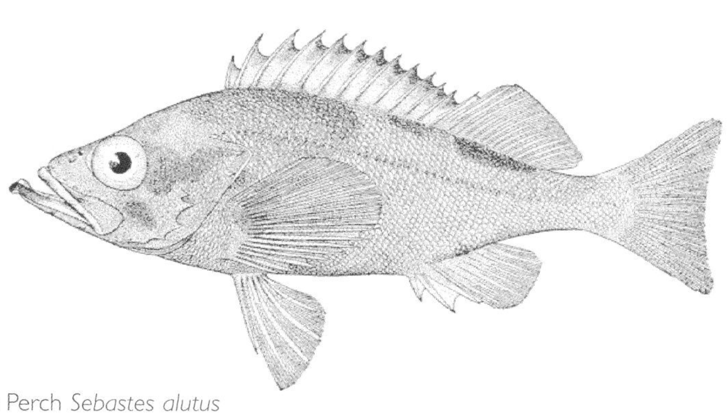

Pacific Ocean Perch *Sebastes alutus*

Credit: Food and Agriculture Organization of the United Nations, Rome, Species Identification and Data Programme, Fisheries Department.

Rockfishes *Sebastes* spp. and *Sebastolobus* spp. (also called sea bass and sea perch) comprise five genera in the family Scorpaenidae. Seven species of rockfish in the genera *Sebastes* and *Sebastolobus* are recorded in catches from the sea bed in the western Bering Sea.

These fish mainly inhabit the waters of the continental slope. The bulk of the catch (86-98%) is made up of **Nothern Sea Bass *Sebastes borealis*** and **Aleutian Bass *Sebastes aleutianus***. Two other species - **Alaskan Rockfish *Sebastolobus alascanus*** and **Long-finned Rockfish *Sebastolobus macrochir*** - are present in catches in relatively small quantities, while the remaining three species recorded in catches in the western Bering Sea are rarely caught and make up only a very small proportion of the catch. One of these species is the **Pacific Ocean Perch *Sebastes alutus***. Directed fishing yields the highest concentrations of Northern Sea Bass, although some of the catch is incidental (Anon., 1998d).

Rockfishes grow slowly and are characterized by their long life spans, which can extend to 30 years. Sexually mature individuals are caught in considerably greater numbers than immature fish and make up the bulk of the commercially exploited stocks. Such long-lived species are extremely vulnerable to poorly controlled fishing and may take a long time (one to three decades) to recover from the effects of over-fishing (KamchatNIRO, pers. comm., 1999).

There was no information available on rockfish stocks.

Fishery

Data on rockfish catches are largely unavailable by species. For the western Bering Sea, catches appear to be highly variable but from 1982-92 they were very low (under 1000 t). For the Pacific Northwest, combined catches of Pacific Ocean Perch for the three nations reporting (Japan, South Korea and the Russian Federation) have varied over the years but in general exhibit a decline (from 6908 t in 1984 to 2440 t in 1998) (see **Annex 2**). (South Korea is a negligible player, reporting a catch of only four tonnes, in 1992 only.) In 1999, FAO statistics reveal that catches of Pacific Ocean Perch for the Pacific Northwest declined further, dropping to 1630 t. FAO catch data does not include records for any other species of rockfish.

Longlines and trawlers are the usual fishing gears for catching rockfishes in the eastern Bering Sea (Witherell, 2000).

Trade

Rockfishes recorded as exported from Kamchatka are not distinguished by species. Japan is the major destination reported, importing on average about 500 t a year for the period 1995-98 (see **Table 16**). In 1997, South Korea entered the market and in 1998 the largest quantity of rockfish exported from Kamchatka that year (467 t) was to that country.

Trends

Trends are unclear for this group of species and analysis is complicated by a lack of species-specific data and the absence of stock information. Since 1995, declines in catches have been observed and interest in this fishery in the western Bering Sea by the majority of Russian fishery organizations has sharply decreased (*Kamchatrybvod*, unpublished, 1999).

Table 16

Rockfish (*Okun*) exports (all forms of fish) from the Kamchatka region

	1995		1996		1997		1998		Jan.-March 1999		Av. annual export '95-98		
	t	USD x1000	t	USD x1000	t	USD x1000	t	USD x1000	t	USD x1000	t	USD x1000	USD/kg
Japan	451.4	1030.4	582.8	2041.1	839.0	1879.6	153.9	173.9	10.0	17.7	506.8	1281.2	2.53
S. Korea	0.0	0.0	0.0	0.0	50.5	39.2	466.7	425.2	0.0	0.0	129.3	116.1	0.90
USA	0.0	0.0	1.1	1.2	0.0	0.0	0.0	0.0	0.0	0.0	0.3	0.3	1.10
TOTAL	**451.4**	**1030.4**	**583.9**	**2042.2**	**889.5**	**1918.8**	**620.6**	**599.1**	**10.0**	**17.7**	**636.3**	**1397.6**	**2.20**

Figure 8

Rockfish exports (all forms of fish) from the Kamchatka region

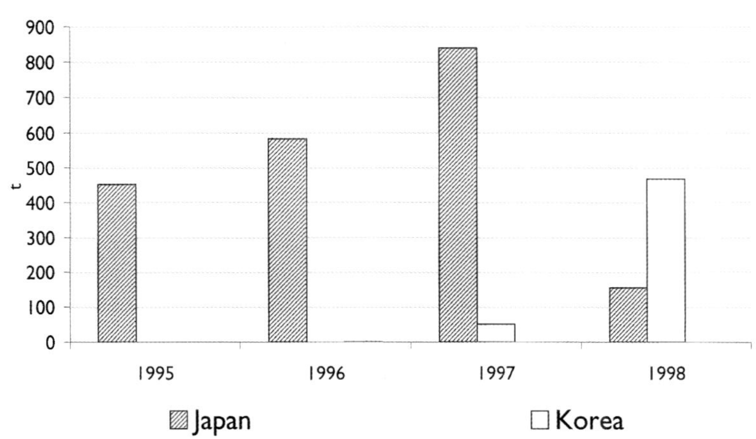

Source: Kamchatka Customs, 1999. **Note**: Korea = South Korea.

Halibuts and flounders (*Paltus*)

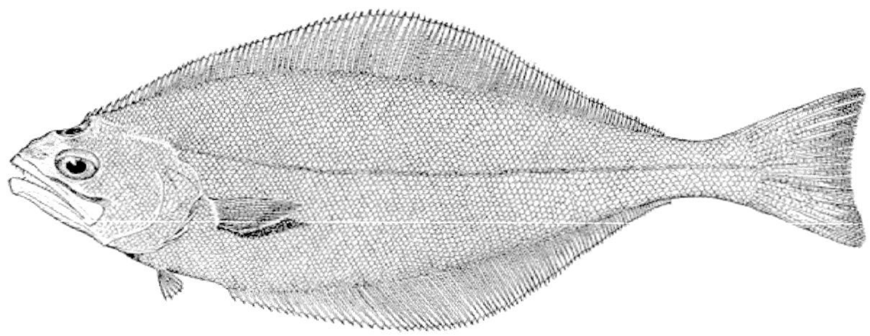

Arrowtooth Flounder *Reinhardtius stomias*

Credit: Food and Agriculture Organization of the United Nations, Rome, Species Identification and Data Programme, Fisheries Department.

The four species of halibut and flounder most often encountered in fisheries in the Bering Sea are **Pacific Halibut *Hippoglossus stenolepis***, **Greenland Halibut *Reinhardtius hippoglossoides***, **Kamchatka Flounder *Reinhardtius evermanni*** and **Arrowtooth Flounder *Reinhardtius stomias***. These species are widely distributed in catches and easily distinguished. Halibuts are encountered along the east coast of Kamchatka in Olyutorsky Bay and the Olyutor-Navarin regions, in the southern parts of Anadyr Bay, and in the central and south-eastern parts of the Bering Sea up to Bristol Bay. Halibuts consume various fish, crabs, shrimps, squid and octopus. Only young halibuts and flounders live in shoals and the mature fish are usually concentrated at depths of 400-750 m. Halibuts and flounders form a constant and noticeable part of the catch from fishing for other ground species (KamchatNIRO staff, pers. comm., 1999).

Stocks

Zaitsev (1996) reports that there are over 20 species of flatfish in the Russian Far East and that the total biomass of these fish is not less than 240 000 t. He further reports that Kamchatka and the north-west Bering Sea are the most abundant regions for flatfish in the region. No specific information about the stock of halibuts and flounders in the western Bering Sea was found.

Fishery

Kamchatka Customs statistics do not distinguish between species but record these type of fish under the general name *paltus*, which means "halibut". Available catch data for *paltus* from the western Bering Sea exhibit a decline, dropping from 5030 t in 1986 to 0.3 t in 1993 (see **Table 6**). Of this group of fishes, only the Kamchatka Flounder appears in FAO statistics for the Pacific Northwest. Two countries report catches of this species to the FAO, the Russian Federation and Japan. Their catches averaged 8800 t a year from 1984 to 1999. They reached a minimum in 1993 (1276 t), but gradually increased after that and reached 10 743 t in 1999. Catch data for Pacific Halibut from the Pacific Northeast show that significant volumes are caught annually by

Table 17

Halibut (*Paltus*) exports (all forms of fish) from the Kamchatka region

	1995		1996		1997		1998		Jan.-March 1999		Av. annual export '95-98		
	t	USD x1000	t	USD x1000	t	USD x1000	t	USD x1000	t	USD x1000	t	USD x1000	USD/kg
China	0.0	0.0	0.0	0.0	19.6	21.8	1.5	3.8	0.0	0.0	5.3	6.4	1.21
Japan	120.8	116.0	748.9	1535.7	1221.0	2517.5	645.2	897.9	48.2	52.2	684.0	1266.8	1.85
Hong Kong	0.0	0.0	0.0	0.0	0.0	0.0	0.0	0.0	7.6	5.2			
S. Korea	160.0	229.8	112.0	184.7	188.4	306.7	1028.0	1711.4	266.6	371.6	372.1	608.2	1.63
Norway	0.0	0.0	0.0	0.0	51.6	129.1	0.0	0.0	0.0	0.0	12.9	32.3	2.50
Ukraine	0.0	0.0	0.0	0.0	0.0	0.0	76.4	21.2	0.0	0.0	19.1	5.3	0.28
USA	26.8	23.7	50.9	114.2	31.0	85.3	40.1	85.7	0.0	0.0	37.2	77.2	2.08
TOTAL	**307.6**	**369.6**	**911.9**	**1834.6**	**1511.7**	**3060.5**	**1791.2**	**2719.9**	**322.4**	**429.0**	**1130.6**	**1996.1**	**1.77**

Figure 9

Halibut exports (all forms of fish) from the Kamchatka region

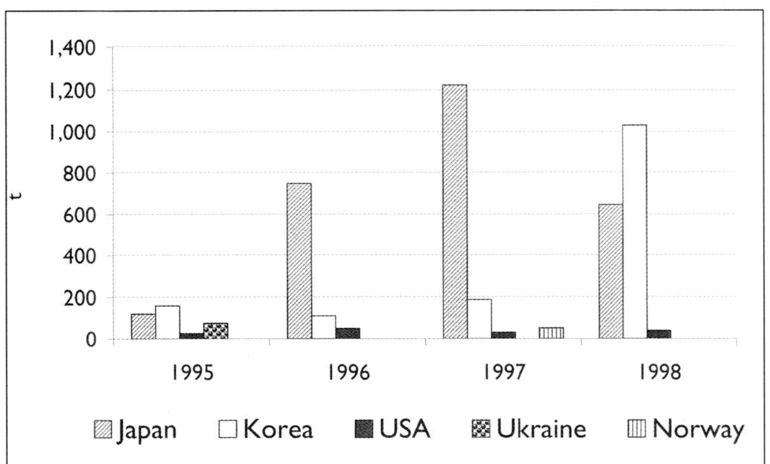

Source: Kamchatka Customs, 1999. **Note**: Korea = South Korea.

Canada and the USA, the average annual catch for the period 1984-99 was 36 412 t, with the maximum in 1999 (81%, or 43 500 t, of which was taken by the US fleet).

Halibuts and flounders have in the past been caught in bottom trawls and bottom nets, in catches with other groundfish. At present bottom trawls are banned and the only target gear allowed is a longline but halibuts and flounders continue to be harvested as by-catch.

Trade and trends

Kamchatka Customs export data indicate a continuous rise in exports, totalling over 5150 t for the period 1995 to the first quarter of 1999. Japan and South Korea were the most important importers (see **Table 17/Figure 9**).

It is difficult to identify trends, when species-specific information is lacking.

Yellowfin Sole and other plaices (*Kambala*)

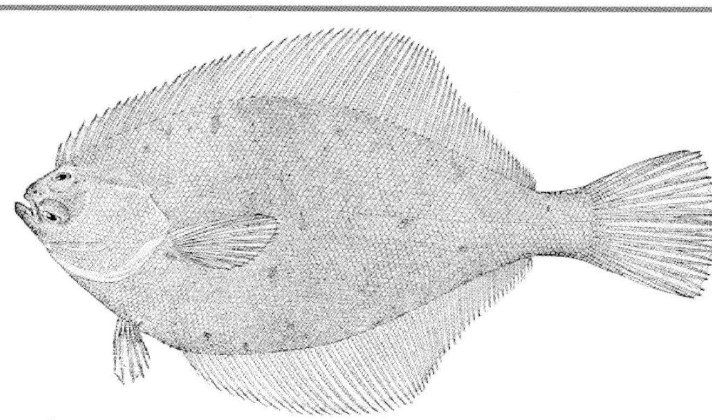

Yellowfin *Sole Limanda aspera*

Credit: Food and Agriculture Organization of the United Nations, Rome, Species Identification and Data Programme, Fisheries Department.

In the bays of the western Bering Sea (Korfo, Karaginsky and Olyutorsky Bays), there are several species of plaice, among which **Yellowfin Sole *Limanda aspera*** is the most prevalent. The majority of Yellowfin Sole in catches measure 23-32 cm and 250-500 g, having reached ages of six to ten years (KamchatNIRO staff, pers. comm., 1999).

Stocks

Yellowfin Sole is subject to fluctuations in numbers between generations, which are characteristic of the species and result in natural changes in the size of stocks.

Fishery

Fishing of plaice is usually carried out by small fleets in the summer to autumn period (*Kamchatrybvod*, unpublished, 1999). Catch data for the western Bering Sea record overall

Table 18
Exports of Yellowfin Sole and other plaices (*Kambala*) (all forms of fish) from the Kamchatka region

	1995		1996		1997		1998		Jan.-March 1999		Av. annual export '95-98		
	t	USD x1000	t	USD x1000	t	USD x1000	t	USD x1000	t	USD x1000	t	USD x1000	USD/kg
China	3508.9	1855.1	1264.5	765.9	2420.0	1071.3	7928.0	2159.4	1958.4	669.1	3780.3	1462.9	0.39
Japan	895.9	1184.3	1006.3	657.8	1107.4	1339.4	709.6	909.1	201.9	210.2	929.8	1022.7	1.10
Hong Kong	0.0	0.0	0.0	0.0	45.4	9.1	119.0	38.1	333.8	67.9	41.1	11.8	0.29
S. Korea	3388.4	1543.5	1695.5	875.8	3163.0	1452.0	2721.8	1097.8	3040.0	1119.5	2742.2	1242.3	0.45
Norway	0.0	0.0	0.0	0.0	0.0	0.0	0.0	0.0	335.8	111.4			
Singapore	107.4	51.6	0.0	0.0	302.2	133.0	1568.5	786.4	0.0	0.0	494.5	242.7	0.49
Taiwan	60.2	33.7	0.0	0.0	0.0	0.0	0.0	0.0	0.0	0.0	15.0	8.4	0.56
Ukraine	295.8	155.4	0.0	0.0	0.0	0.0	3442.9	999.4	0.0	0.0	934.7	288.7	0.31
USA	0.0	0.0	0.0	0.0	59.0	27.4	945.7	238.8	0.0	0.0	251.2	66.5	0.27
TOTAL	**8256.5**	**4823.6**	**3966.3**	**2299.5**	**7097.0**	**4032.2**	**17 435.6**	**6229.0**	**5869.9**	**2178.1**	**9188.9**	**4346.1**	**0.47**

Source: Kamchatka Customs, 1999.

Figure 10
Exports of Yellowfin Sole and other plaices (all forms of fish) from the Kamchatka region

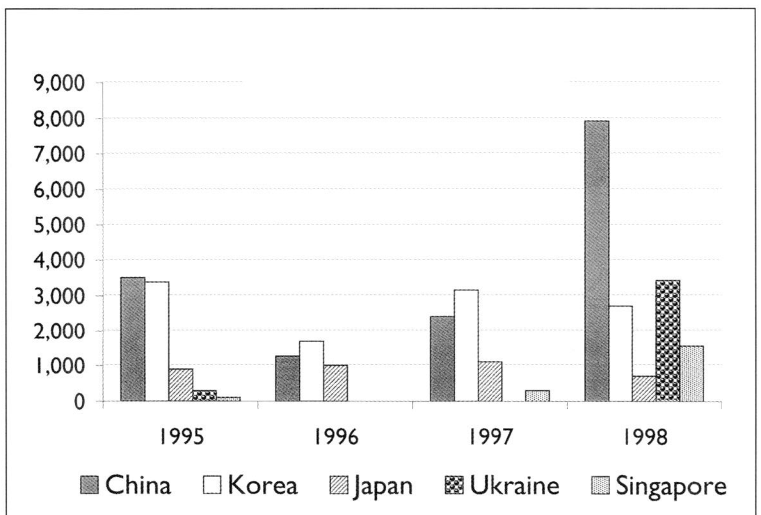

Source: Kamchatka Customs, 1999.
Note: Korea = South Korea.

catches of this group of fishes (Pleuronectidae) and are not species-specific (see **Table 6**). For the wider Pacific Northwest, catches of Yellowfin Sole have not been reported since 1986, while in the Pacific Northeast the only reported fishery was in the USA, where the annual catch averaged around 119 750 t, 1984-99. From 1997 to 1999, the catch dropped by 62% from 149 300 t to 56 830 t.

Credit: Alexey Vaisman, TRAFFIC Europe-Russia

Young flatfish are observed among pollack and cod in the hold of an impounded trawler in the western Bering Sea in July 1999.

Bottom-nets and seines are used to catch Yellowfin Sole. It is one of the most common species to occur as by-catch in bottom gears also (*Kamchatrybvod*, unpublished, 1999).

Trade and trends

Trade reported by Kamchatka Customs for 1995-98 is shown in **Table 18**. It is not possible to identify trends without species-specific information.

Shrimps

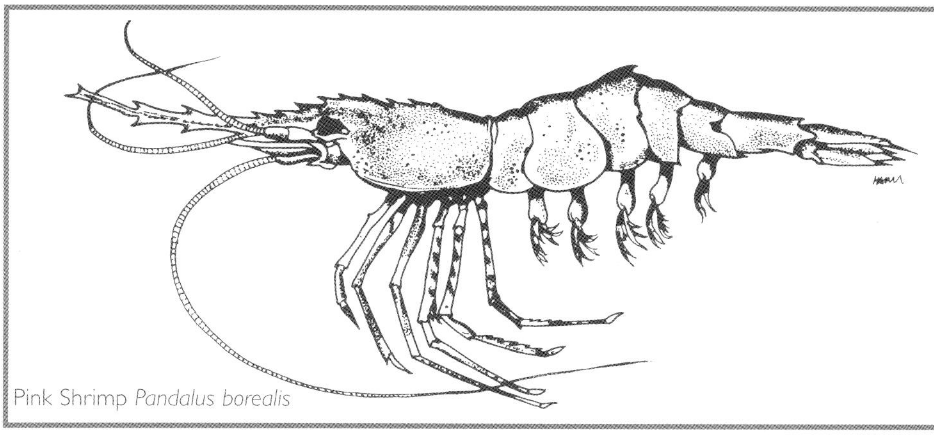

Pink Shrimp *Pandalus borealis*

Credit: Food and Agriculture Organization of the United Nations, Rome, Species Identification and Data Programme, Fisheries Department.

The bulk of the shrimp biomass in the Bering Sea consists of two species, **Pink Shrimp *Pandalus borealis***, which occurs in the greatest numbers in the eastern part of the sea, and **Humpy Shrimp *Pandalus goniurus***, which is more abundant in the western part. At times, these species form very large accumulations, the former around the Pribilof Islands and the latter in the Navarin region, including the southern part of Anadyr Bay. Both the Pink Shrimp and the Humpy Shrimp are small and are marketed as salad or cocktail shrimps (Watson, 1994).

Stock

The biomass of Humpy Shrimp in the western Bering Sea was estimated to be 600 000 to 725 000 t in the 1970s (Shuntov *et al.*, 1995). In the 1980s the estimated stock declined three- to four-fold, to about 100 000 to 200 000 t (Safronov, *in litt.*,1999). According to the National Academy of Sciences (NAS) in the USA, little is known about shrimp stocks in the western Bering Sea (Anon., 1996a), yet VNIRO scientists report that deepwater shrimp stocks in the Russian part of the Bering Sea and the areas off the eastern Sakhalin coast and the south-western coast of Kamchatka are underexploited (Anon., 2000m). Both the Pink and Humpy Shrimp are subject to large inter-annual fluctuations in number (Shuntov *et al.*, 1995).

Fishery

The shrimp fishery in the region appears to have expanded after the invention of the mechanical peeler in 1958 (Watson, 1994). Exploitation of shrimps in the region began around the Pribilof Islands in the early 1960s and was undertaken by Japanese fleets. Towards the late 1960s and early 1970s, both Japanese and Russian fleets were fishing for shrimp in Anadyr Bay and the northern and central parts of the Bering Sea. This large-scale fishing caused the shrimp biomass to diminish dramatically in these regions. By 1967, the fishery was regarded as being "inconsequential" and while knowledge on the subject is sparse, it is thought that stocks in the western Bering Sea have not recovered from earlier over-fishing (Balsiger, 1981 in Anon., 1996a). Data from the State Committee of Fisheries (Moscow) on catches in the western Bering Sea only report catches for seven separate years for the period 1965-87. These total 5419 t, with 3500 t of this quantity caught in a single year - 1987 (Anon., 1996a).

The reduction in shrimp biomass may not be attributable solely to over-exploitation, as predation and changes in sea conditions can also have a great influence on shrimp numbers (Safronov, *in litt.*,1999).

For the Pacific Northwest as a whole, the Russian Federation has reported steady catches of *Pandalus* shrimp every year for the period 1984-97, the average annual catch being 1919 t. From 1997-98, there was a 100% increase in catch for the Russian Federation in this area, but there are no Russian catches were reported to FAO for the area in 1999 (see **Annex 2**).

No information is available on fishing gear.

Trade and trends

No exports are reported by Kamchatka Customs. Shrimp catches in recent years have not been sufficiently abundant to fill quotas: it is possible that shrimp stocks remain in a state of decline.

Commander Squid

The **Commander Squid *Berryteuthis magister magister*** is the main commercially exploited squid species in the western Bering Sea. Other species of squid with commercial potential also occur in Russian waters, in particular **Short-finned Squid** (also known as Red Flying Squid) ***Ommastrephes bartramii*** and **Japanese Flying Squid *Todarodes pacificus pacificus*** (Anon., 1999h). The Commander Squid forms dense concentrations around the Commander Islands in the western Bering Sea. It also congregates around the Pribilof Islands in the eastern Bering Sea and the Kuril Islands, which lie south of Kamchatka, stretching towards Japan (Anon., 1999c; Day, 2000).

Stocks

The Commander Squid stock is estimated to amount to 400 000 to 500 000 t for the area comprising the Sea of Japan, the Sea of Okhotsk and the Bering Sea (Anon., 1999c).

Fishery

At present, it is estimated that squid stocks are being exploited at only 50-60% of their potential. The catch is primarily destined for domestic markets, as the method of fishing, by trawl, apparently damages the product and leaves it unacceptable for export (Anon., 1999h). According to FAO data, the Russian share of catches of various squids from the Pacific Northwest, 1984-99 was about 40% of the total, averaging 43 800 t annually. Russian catches increased from 15 750 t in 1994 to 54 760 t in 1999. The catch of squid from the Pacific Northeast was only 1500 t in 1999, 85% of which was taken by South Korea, the rest by the USA.

It is reported that no special permits are required to catch a squid, hence there is no quota (Anon., 1999d).

Trade

As most of the catch of squid is for domestic consumption, reported exports of squid from Kamchatka appear to occur on an irregular basis, with no exports reported for 1996 and 1997, nor for the first three months of 1999 (see **Table 19**).

Table 19
Squid (*Kalmar*) exports (all forms) from the Kamchatka region

	1995		1996		1997		1998		Jan. -Mar. 1999		Average annual export '95-98		
	t	USD x 1000	t	USD x 1000	t	USD x 1000	t	USD x 1000	t	USD x 1000	t	USD x 1000	USD /kg
China	172.4	69.0	0.0	0.0	0.0	0.0	589.1	430.7	0.0	0.0	190.4	124.9	0.66
S. Korea	0.0	0.0	0.0	0.0	0.0	0.0	7.5	4.5	0.0	0.0	1.9	1.1	0.60
Ukraine	0.0	0.0	0.0	0.0	0.0	0.0	445.1	219.4	0.0	0.0	111.3	54.8	0.49
USA	0.0	0.0	0.0	0.0	0.0	0.0	0.1	0.2	0.0	0.0	0.025	0.040	1.60
TOTAL	**172.4**	**69.0**	**0.0**	**0.0**	**0.0**	**0.0**	**1041.8**	**654.7**	**0.0**	**0.0**	**303.5**	**180.9**	**0.60**

Source: Kamchatka Customs, 1999.

Trends

There is the potential to increase exploitation and this may well happen if other resources dwindle. The market for squid is growing, particularly in the USA and China, but these and other markets are already supplied by numerous species of squid (such as California Loligo Squid *Loligo* sp. and Argentine Shortfin Squid *Illex argentinus*) (Anon., 2000n; Anon., 2000o).

EXPORTS OF FISHERIES PRODUCTS REPORTED FROM KAMCHATKA

Export data compiled by Kamchatka Customs are presented in **Tables 20**, **21** and **22**. At present, the main importers of fish and other marine resources from Kamchatka, in terms of volume, are South Korea, Japan, China and, to a lesser extent, Singapore (see **Table 21**).

Table 20 indicates that a jump in exports occurred in 1997. In the latter part of the 1990s, (see ***Overview of the fishery in the 1990s***), there was a move within the Russian fishing industry of the Bering Sea to fill foreign demand, rather than to cater for domestic markets. The sharp increase in exports in 1997 may be explained by this shift - from shore-based processing for the domestic market, to processing at sea for export.

Table 20

Annual volumes of the major fisheries products (all forms) exported through Kamchatka Customs

	1995		1996		1997		1998		Jan. - Mar. 1999	
	t	USD	t	USD	t	USD	t	USD	t	USD
Alaska Pollack	2538	941 000	2360	1 627 000	46 557	25 537 000	78 523	52 337 000	27 521	19 705 000
Crab	5132	62 438 000	2959	25 715 000	4788	34 198 000	3891	28 782 000	589	4 160 000
Pacific Cod	7038	4 288 000	9671	6 174 000	16 483	10 962 000	18 566	11 167 000	8669	5 477 000
Saffron Cod	1916	1 205 000	38	19 000	163.3	286 000	4615	1 524 000	1848	349 000
Herring	8166	4 081 000	6624	4 203 000	17 341	7 703 000	21 530	5 174 000	4130	1 584 000
Rockfish	451	1 030 000	584	2 042 000	889	1 919 000	621	599 000	10	18 000
Halibut	308	370,000	912	1 835 000	1512	3 061 000	1791	2 720 000	322	429 000
Plaice	8196	4 790 000	3966	2 300 000	7097	4 032 000	17 436	6 229 000	5870	2 067 000
Squid	172	69 000	0	0	0	0	1085	672 000	0	0
Salmon	3103	11 303 000	7233	24 354 000	7155	15 610 000	37 259	39 207 000	204	145 000
Salmon roe	345	3 823 000	163	1 248 000	213	1 159 000	2221	7 421 000	3	16 000
Sea urchins	71	136,000	10	12 000	0	0	0	0	0	0
Scallop	226	1 682 000	139	812 000	384	2 917 000	616	4 391 000	184	1 208 000
Terpug[1]	1703	832 000	1729	865 000	1567	857 000	3204	975 000	141	86 000
TOTAL	**37 414**	**96 155 000**	**34 264**	**70 606 000**	**101 097**	**105 211 000**	**201 388**	**165 394 000**	**49 491**	**35 244 000**

Source: Kamchatka Customs statistics, 1999.

[1] Greenling or Atka mackerel = *Pleurogrammus azonus*

Table 21

Annual fisheries exports from Kamchatka, by country of import, 1995 to March 1999

Country	1995		1996		1997		1998		Jan - Mar 1999		TOTAL	
	t	USD x1000	t	USD x1000	t	USD x1000	t	USD x1000	t	USD x1000	t	USD x1000
Canada	0	0	0	0	43	334	659	277	0	0	702	611
China	11 277	5628	5544	3627	29 267	14 223	44 709	17 142	10 381	4406	101 178	45 026
Japan	15 823	87 358	19 136	61 495	31 929	78 473	34 509	65 583	1040	3875	102 438	296 783
Hong Kong	0	0	0	0	425	152	4652	3104	2177	369	7254	3625
Germany	0	0	687	915	6474	5059	4389	6402	0	0	11 551	12 377
Korea	9886	6816	10 697	9577	16 676	9546	58 332	48 177	17 236	10 674	112 826	84 789
Singapore	0	0	0	0	16 205	8625	9819	6001	5902	3931	31 926	18 558
Ukraine	3027	1472	0	0	0	0	15 704	4571	2129	669	20 859	6713
USA	321	7864	1009	2846	1349	4528	17 414	19 353	3620	7232	23 712	41 821
Virgin Is.	0	0	0	0	0	0	10 338	6278	6195	4064	16 532	10 341
TOTAL	**40 333**	**109 137**	**37 072**	**78 460**	**102 369**	**120 940**	**200 524**	**176 887**	**48 680**	**35 220**	**428 978**	**520 644**

Source: Kamchatka Customs statistics, 1999.

Table 22

Export prices (USD/kg) for selected fisheries products (all forms) exported from the Kamchatka region, 1995-99.

	1995	1996	1997	1998	1999
Crab	12.17	8.68	7.14	7.40	7.06
Salmon roe	11.08	7.68	5.44	3.34	5.04
Scallop	7.44	5.83	7.66	7.13	6.57
Salmon	3.64	3.37	2.18	1.05	0.72
Pacific Ocean Perch	2.28	3.50	2.16	0.97	1.77
Halibut	1.20	2.01	2.02	1.52	1.33
Cod	0.67	0.74	0.62	0.54	0.63
Saffron Cod	0.64	0.50	0.33	0.33	0.19
Plaice	0.58	0.58	0.57	0.36	0.35
Herring	0.50	0.63	0.44	0.24	0.38
Atka Mackerel	0.49	0.50	0.55	0.30	0.61
Squid	0.40			0.62	
Alaska Pollack	0.37	0.69	0.55	0.67	0.72

Source: Kamchatka Customs statistics, 1999.

Of marine products exported shown in **Table 22**, crabs were the most valuable per kilogramme, but from 1997 Alaska Pollack was nearly as valuable export by virtue of the large volumes shipped (see **Tables 20** and **22**). This much explains the fierce competition surrounding harvest of crabs and pollack in the Russian part of the Bering Sea. **Table 22** also shows that the value of various species, in terms of export price per kilogramme, has steadily declined for many species.

FISHERIES VIOLATIONS IN THE RUSSIAN FAR EAST, WITH SPECIFIC REFERENCE TO THE BERING SEA

Data from law enforcement bodies, the offices of the public prosecutor, the Ministry of Foreign Affairs and from Russian and other media provide evidence for the case that Russia's fishing industry has turned into a powerful black-market sector, representing huge losses for the country's economy. The State's weak executive powers and corruption in many of the inspectorates have allowed this rise in crime and illegal harvest processing and trade have taken on an increasingly organized character from year to year since the early 1990s.

While investigating the fisheries of the western Bering Sea, researchers found illegal activities at virtually all levels of the industry, in nearly all seafood markets and especially where pollack was concerned. The illegal harvesting of marine products has become the main objective for many players involved in the Bering Sea fisheries sector. In 1995, as joint ventures for harvesting and processing fish dwindled to two and 11, respectively, the number involved in trade rose from 28 in 1993, to 37 in 1995 (see ***Overview of fishery in the 1990s***; TRAFFIC survey data) and an outflow of capital from the region took place as companies were able to escape Customs and other checks and associated duties, facilitated by poor international trade controls.

According to the Russian State Committee on Statistics, the country's foreign-trade turnover in fish products has shown a negative balance since 1995 and the estimates for the value of trade lost through illegal activity range from one to five billion US dollars per year (see **Table 23**).

Table 23
Estimates of losses (USD) to illegal fishing activity

Period	Source	Amount	What? from where?	Date of estimate	Reference
Annual	"Specialists"	2 billion	Losses in Far Eastern waters	1994	Reznik and Ostrovskaya, 1994
1996	Director of the Federal Border Service	4 billion	Bioresources from Russian waters	Oct. 1997	Ruchkin, 1997
Annual	Office of the Prosecutor General	2 billion	Marine products from marine areas	Aug. 1997	Mitin, 1997
1990-97	Director of the Federal Border Service	5 billion/year	Revenue lost from illegal export of sea products	Oct. 1997	Ruchkin, 1997
Annual	"Experts"	2-5 billion	Losses due to inadequate oversight of harvesting and export of marine products	1997	Plotnikov, 1997
?	President Vladimir Putin	2.5 billion	Unaccounted fish exports	July 2000	Myles, 2000
Annual	Chairman of the State Fisheries Committee, Yuri Sinelnik	700 million to 1 billion	Illegal fish exports from all basins of the country	Sept. 2000	Anon., 2000p

Table 24
Violations of the *Fisheries Act* detected in the western Bering Sea, 1993-98

Violations	1993	1994	1995	1996	1997	1998	1999
Failure to record accurate data in vessel documentation	46%	46%	40%				
Distortion of data in vessel documentation		33%		38%	51%	>50%	
Unauthorised dumping of waste	23%			18%			
Fishing in prohibited areas			40%	18%			
Other	31%			26%	49%		
Nationality of vessels involved in offences							
Russian vessels	13	54	-	51	33	11	152
Foreign vessels	6	2	-	1	4	4	57
Total	**19**	**56**	**12**	**52**	**37**	**15**	**209**

Source: Kamchatka Basin Regional Fisheries Inspectorate, 1999.
Note: Data for 1995 are not available in their entirety.

Table 25

Selected examples of illegal activities detected in the fisheries sector of the Russian Far East

Date	Name and national registration of the	Nature of the infraction	Penalty imposed	Quantity of products seized	Reference
1995	*Putina* '95 resulted in fines imposed on 123 Russian vessels and two foreign vessels	Unknown	RUR15 million imposed on 123 Russian vessels; USD40 000 on two foreign vessels	Unknown	Pautzke, 1997
1996	*Pacific Kim*: (S. Korea)	Catch and transport of crab without a permit	USD100 000	64 t of crab	SMIS, 1999
1996	*Putina* '96 resulted in 2195 violations	2195 violations (2152 for poaching)	More than RUR2 billion	425 vessels	Pautzke, 1997
1997	Steamer *Byakin* DVMP (Russia)	Catch and transport of fish and crab products without original permit or licence for fishing activity	USD100 000	150 t of crab products, and 550 t of pollack products	SMIS, 1999
1997	Trawler *im. 61 kommunara*: (Ukraine)	Transfer and transport of fish products without a permit	Unknown	3800 t of mixed fish products	SMIS, 1999
1997	*Putina* '97, stage 1 resulted in 294 violations	Unknown	RUR7 billion	RUR6.8 billion worth of fish and sea products	Pautzke, 1997
1998	*Tai Ping* (China)	Illegal catch of 200 t of Alaska Pollack in the Sea of Okhotsk	Unknown	Unknown	Anon., 1998g
1998	*Fire Sea* Marina Ich (Russia)	Misrepresentation of l volume of the catch. Illegal catch and attempted export of crab without relevant permits	Captain Sharaban given four years in prison	Unknown	SMIS, 1999
1998	*Putina* '98, stage 1 resulted in 157 violations	Unknown	Unknown	323 t of illegal fish products	*Kamchatrybvod* staff, *in litt.*, December 1999
1999	*Kastrikum* Merkurii marine convoy, Marina Ich (Russia)	Misrepresentation of the volume of the catch. Attempted export of 100 t of Kamchatka King Crab, concealed underneath Golden King Crab	Unknown	100 t of Kamchatka King Crab	SMIS, 1999
1999	*Yasnomorskii* Vostoktransservis (Russia)	Catch and transport without a permit, withholding information, discrepancies between actual and permitted crab catch	Fine of 2000 minimum statutory wages	Vessel and 23 t of crab products	SMIS, 1999
2000	8 Japanese vessels	Poaching in Russian waters	Total of USD2.5 million	Unknown	Anon., 2000q
2000	Japanese trawler *Daikotu maru-11*	Poaching in Sea of Okhotsk	USD670 000	350 t of fish	Anon., 2000r
2000	Japanese trawler *Seidgu maru-21*	Poaching in Russian waters	USD506 000	350 t of fish	Anon., 2000r
2000	Japanese fishing boat *Dantei Maru-5*	Salmon poaching in the Commander Islands; 20 t of illegally caught salmon on board	USD200 000	Unknown	
2000	Japanese driftnetter *Umitaka 15*	Fishing in an illegal zone (a salmon run); using driftnets double the allowed length; using unmarked buoys	USD33 000	Unknown	Fossbakk, 2000a
2000	Two Japanese fishing boats, the *Koshin Maru-1* and the *Anjo Maru-18*	Leaving Russia's free economic zone without undergoing the required mandatory inspection	USD15 385 each	Unknown	Fossbakk, 2000b
2000	Three Russian fishing boats *Udyl, Bor* and *STN-18*	Poaching more than 15 t of crab	Licences suspended, pending investigation	Unknown	Fossbakk, 2000b
2000	Japanese fishing boat *Miyadzima Maru*	Fishing without a permit within the Russian EEZ in the Sea of Japan	The captain was fined ca. USD7500 in minimum statutory wages and ca. USD33 600 in damages (to marine stocks)	Fishing boat, fishing gear, and 30 t of squid	Anon., 2000s

* **Note**: DVMP = *Dal'nevostochnoe Morskoe Parokhodstvo*, a Far Eastern Russian shipping company

This amount is two to 10 times the value of all exports declared to the Regional Kamchatka Customs Branch between 1995 and 1998 (see **Table 21**).

The types of violations are numerous: **Table 24** shows the most common violations recorded by the fisheries inspectorate in Kamchatka, while **Table 25** contains a summary of major violations discovered over a six-year period within the fisheries sector in the whole of the Russian Far East. In 1999, control patrols were reportedly intensified, which may account for the rise in reported infractions in that year.

A high proportion of offences are typically committed in the summer months. According to the Kamchatka Basin Regional Fisheries Inspectorate, percentages for 1993-97 were the following:

	Summer	*Autumn*	*Winter*	*Spring*
1993:	ca. 100%			
1994:	55%	17%	0%	28%
1995:	no data	no data	no data	no data
1996:	4%	61%	4%	31%
1997:	76%	15%	3%	6%

The higher frequency of infractions occurring in summer coincides with the higher level of fishing intensity during this period (see **Table 3**). In the early summer, female pollack and salmons are sought for valuable roe as they migrate to coastal waters. Also relevant are the milder weather conditions of summer, when the days are often clear and warm, compared to misty and freezing in the winter. This not only permits smaller fishing vessels with less sophisticated equipment to fish, but also allows small vessels from the fisheries enforcement agencies to undertake their inspection missions (TRAFFIC survey data).

Information compiled on the violations referred to in **Table 24**, as well as information gathered on fisheries violations occurring in the Russian Far East in general is presented below. The types of violation are listed in order of priority (frequency of incidence).

Types of violation

Falsifying documents

According to Russian law, ships' logs must include reports of all fishing activities, specification of fishing gear used, the time of its use, the volume of the catch and its composition by species and size of specimens caught. During this study, interviews with fisheries inspectors revealed that almost all vessels had two fishing logs: an official log for the inspectors and a "confidential" log for the owner (TRAFFIC survey data). Such falsification of documents appears to be used widely in an attempt to conceal a range of illegal operations and is therefore a broad category of violation used to facilitate a variety of others, including under-reporting harvest; recording of a false vessel location; illegal acquisition of fishing quotas and illegal offloading of fish. Some notable examples of violations committed with the aid of false papers are described in the following pages.

Exceeding permitted catch volumes and associated unauthorized sale and export of fisheries products

Altering documentation so that the amount of product declared to be on board does not exceed the permitted quantity is the easiest and most widely used method of concealing surpluses. In other words, the offences of falsifying documents and exceeding permitted catch volumes are closely interlinked. Based on findings of research conducted for this report, far from keeping within legal limits, fishers in the western Bering Sea attempted to catch as much fish as possible. The Kamchatka Directorate of Federal Security Service notes that "based on information available to us, many crew members sell fish and products produced at sea for cash, either with the consent of the vessel's owner or on their own initiative. What is sold for cash is usually the surplus, i.e. fish products that are not accounted for, and which was created by manipulating the catch allowances, or by taking a larger quantity of raw products than shown in the acceptance/delivery receipts and production logs." Illegally harvested products are transferred from one vessel to another. This is called the "captain's trade".

In some instances, TRAFFIC learned, logs are maintained in such a way so as to be adjusted easily in the event of an on-board inspection. For example, if a vessel has a quota for 100 t of Alaska Pollack and catches this amount in five separate hauls, all five operations will be recorded but with only 10 t per haul entered for the first four catches and nothing for the last one. In this way, 40 t of fish is registered on board a ship that in fact has 100 t. When arriving in the fishing port, the ship unloads its catch of 40 t to the official recipients of the cargo, sells the remaining 60 t "under-the-table", and sets off for its next trip. If the ship is inspected when it arrives in port, the captain will have a few seconds to fill the empty column with the necessary entry of 60 t and the total amount of fish on board corresponds to the ship's catch records. It should be noted that on board inspections occur rarely, only three to four times a year according to Federal Border Service chiefs and inspectors (TRAFFIC survey data).

These tactics for log manipulation, which allow a smaller portion of the catch than is actually on board to be registered while the remainder is distributed between customers who pay in "black bread" - unrecorded cash deals - are sometimes referred to as the *Kuril Hokkutensen* method. This is the case when the ship in question is a Russian vessel delivering illegal catches directly to the Russian Federation. Where a non-Russian ship is involved, methods may differ from these described, in so far as laws governing documentation requirements may vary from Russian laws. However, it is often difficult to distinguish between Russian and non-Russian operations since many foreign fishing ships in the Russian EEZ are rented by Russian firms and operate under the permitted allowances and quotas of these firms.

Those carrying illegal catches are usually in a hurry to pass them on and they are therefore often sold very cheaply. They may be sold from ship to ship for foreign cash. In the words of fishermen, "we give them fish and they give us a load of bucks". In August 1993, for example, the trawlers *Kizer*, *Moskovskaya Olympiada* and *Mys Orekhova* belonging to the joint-stock company *Okeanrybflot* produced 50 t of undocumented pollack roe and transferred it to the freezer trawler *Tesey*. The black market product was then sold by the *Tesey* in the port of Pusan

(South Korea) for USD220 000, although based on 1993 prices it would have been possible to get USD500 000 for the same roe (Anon., 1996b).

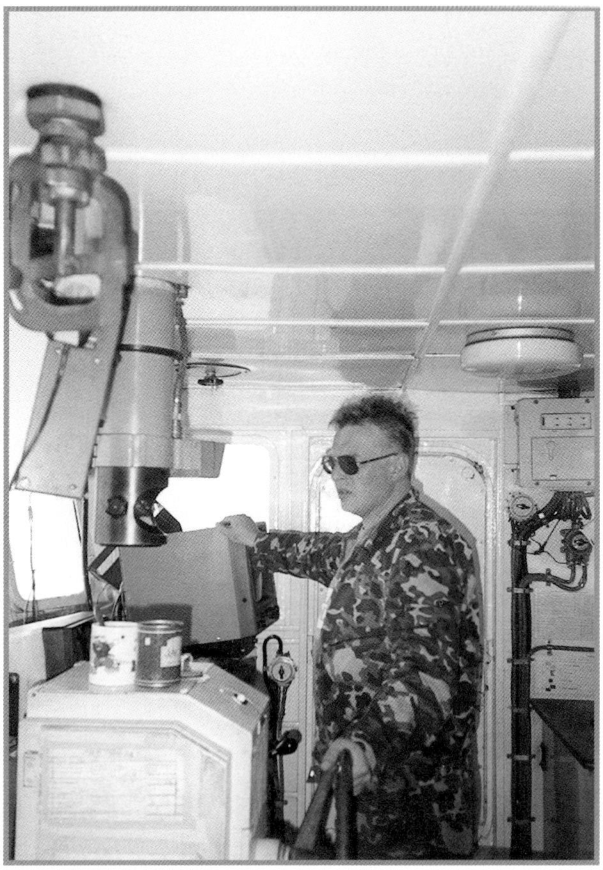
Credit: Alexey Vaisman, TRAFFIC Europe-Russia

Chief of the inspection ship the *Ingeneer Martynov*, on watch

A similar report of under-pricing concerned 146 Russian ships entering the Japanese port of Yokohama over a two-week period in the late 1990s, selling crab to Japanese traders at USD4.50/kg, while the official minimum price at the same port for the product was USD10/kg.

The identity of offenders apprehended by fisheries inspectors in the Kamchatka region has not been obtained, but researchers received information that in 1994 just five joint-stock companies committed between three and 11 violations each, at least. In 1998, more than 70% of the offences committed by Russian fishing vessels involved ships of just two companies, (with shares of 45% and 27%).

A comparison of international trade data recorded by the Russian Federation on the one hand, and by importing countries on the other, provides some indication of the scale of unauthorized exports. Export values recorded by the Kamchatka Regional Directorate of Federal Security of the Russian Federation (including high-value products such as crab, shrimp and salmon) are compared with supposedly corresponding imports by Japan for 1995-97 in **Table 26**.

Table 26

Comparison of fishery commodities' values as reported by importers *versus* exporters

	Imports based on reports by Japanese Customs (million USD and billion JPY)	Exports based on Russian fishery statistics (million USD and billion JPY)	Discrepancy between Japanese and Russian data
1995	USD303.6/JPY32.3	USD108.1/JPY11.5	-64.4%
1996	USD419.2/JPY44.6	USD78/JPY8.3	-81.4%
1997	USD441.7/JPY47.0	USD112.8/JPY12.0	-74.5%

Sources: Kamchatka Regional Directorate Federal Security, unpublished; Japanese Customs statistics.

Note: All types of crab, shrimp and salmon are included. JPY = Japanese Yen

As can be seen from **Table 26**, for three consecutive years the value of imports reported by Japanese Customs far exceeded the value of exports reported by Russian enterprises, indicating either that Japan exaggerated its imports or that exports from the Russian Federation were under-reported. In the latter case, Japanese-reported imports of Russian seafood are likely to have been made up partly by illegal catches, which traders chose to export through unofficial routes to avoid detection by the authorities.

There have been reports of catches sold to Russian carrier vessels bound for ports in Japan, South Korea, China, the USA and Canada - another means of unofficial export. According to staff at *Kamchatrybvod*, the Special Marine Inspection Service and the Far-Eastern Customs Department, the South Korean port of Pusan is especially favoured by fish smugglers. Russian vessels arriving in Pusan with unorthodox loads take the chance that no-one will go to the length of unloading and reloading several thousand tonnes of fish in order to check correlation with the ship's documents. The reasoning is that, with a ship packed to capacity, with no excess space for shifting parts of the catch to allow checking of other fish underneath, an inspection would only be possible with an empty refrigerated vessel on hand. As this would be costly and often impractical, inspections of such fully loaded ships are unlikely.

Money laundering to obscure illegal catches/profits

In contrast to the use of under-pricing to facilitate sale of illegally catches, prices may be deliberately elevated in Customs documentation to conceal profits from illegal harvests. For example, during the period 1995-99, the prices recorded by Russian vessels for Alaskan Pollack products far exceeded accepted market prices listed in trade publications such as *Kommercheskij byulleten* (Commercial Bulletin) and *Groundfish* at that time. Specifically, prices declared to

Table 27
Price comparisons for Alaska Pollack products (USD/kg)

Type of product	1995	1996	1997	1998	1999
Frozen whole Alaska Pollack	2.7	2.7	5	2.5	3.3
Headless frozen Alaska Pollack	3.8-4.9	2.1-2.6 1.38*	1.3-2.5 0.82-1.54**	1.7-2.4	1.5-2.5 1.05*
Frozen Alaska Pollack in blocks		2.5 1.80-2.00*	1.45-1.80* 1.16-1.70**	2.10-3.20*	1.55-3.25*
Alaska Pollack fillets	0.8	0.7-0.9 2.15-2.30*	0.8 1.60-2.37* 1.45-1.80**	0.6-1.8 1.85-3.28*	0.9 1.86-2.81*
Alaska Pollack paste		1.1-1.6 1.10*	2 0.75-1.90* 0.85**	1.5-2.5 1.40-1.60*	1.8 1.20*
Fresh Alaska Pollack			10		6.4

Sources: State Customs declarations; * *Groundfish;* ** *Kommercheskij Byulleten.*

Customs for headed Alaska Pollack, blocks of Alaska Pollack and pollack paste were respectively 1.7, 1.2, and 1.3 times higher than the accepted market prices (**Table 27**). Filleted fish was an exception, for which the declared price was roughly 2.4 times less than the market price.

The theory that these falsely high prices indicated the need to legalize money earned from illegal catches in excess of quotas was supported by interviewees in the region. Albeit probably coincidentally, the average proportion of 50% by which the prices declared to Customs exceeded accepted market prices corresponds to the proportion by which catches of Alaska Pollack are estimated to have exceeded legal quotas (see **Alaska Pollack, Fishery**: TRAFFIC survey data).

Re-selling of quotas

Quotas are often illegally re-sold. Two of the most common ways in which this is done are described below.

- A quota recipient enters into a deal with a powerful foreign or domestic company. The latter provides fuel and vessel repairs where needed and in return is allowed to buy eventual catches from the quota holder at prices lower than market level (TRAFFIC survey data).

- A quota holder who does not own a boat and/or does not have the capacity to process catches, enters into a contract with a company having these facilities. The role of the quota holder in this case is to supply the quota document, in return for a share of profits. The percentage of the profits is usually agreed in the contract (TRAFFIC survey data).

Researchers learned of a specific example of quota re-sale, involving a director of a joint-stock company and a director of a fishing co-operative, both based in Kamchatka. The scheme was dependent on other types of fraudulent activity, such as document falsification. Foreign as well as Russian businessmen and employees of law enforcement bodies were implicated in the affair (Anon., 1998e).

Fishing in prohibited regions

This is a frequent type of violation committed by organized groups of vessels whereby fishing is carried out in areas which have been closed, for example to allow spawning to take place. According to available unofficial data from 1998, as many as 80 vessels operated in areas where fishing was prohibited at some time. Selected examples of this type of infringement are given below.

- In an instance in 1999, a Japanese ship was detained in a marine nature reserve off the Commander Islands, where fishing is strictly prohibited. An observer was on board at the time.

- In 1997, two vessels of a Kamchatka joint-stock company were engaged in crabbing in the Cape Navarin area. One received an order while bound for port in the USA (at Dutch Harbour) to present for inspection by the fishery protection vessel *Manchzhur*. Seeking to avoid the checks, the captain of the vessel issued false information about its activities and locations and did not respond to the request for an inspection for almost a day. When the vessel was eventually apprehended and inspected, it was found to have fished in unauthorised areas (Anon., 1998f).

Unauthorized switching of the species targeted

It is common practice for captains at sea with a fishing licence for a relatively low value species, such as flounder, to start fishing for higher value species, such as salmon, without authorization. These illegal catches are concealed by placing a layer of the licensed, lower value species on top of the unauthorized catch.

Credit: Alexey Vaisman, TRAFFIC Europe-Russia

A fisheries inspector is checking the size of fish and the percentage of various species among the catch, which consists mostly of pollack, with flatfish and some valuable illegally caught crabs.

Use or presence of prohibited fishing gear

A frequent violation when fishing for pollack is the use of bottom trawls to catch the more valuable spawning fish and their roe. It was noted that in the 1998 pollack season, poachers made wide use of bottom trawling, which is prohibited throughout the Bering Sea, resulting in significant catches of pollack fry and other young fish which were sent for processing or, at worst, disposed of.

Also prohibited, but widespread, in the pollack fishery is the hauling of a bottom seine directly on the sea bed to catch crabs. In the case of an inspection vessel appearing, it can always be claimed that the navigator in charge made a minor error. The opposite also occurs, where bottom seines are hauled in shallow waters to catch a shoal of salmon instead of pollack. The captain of one fish-processing ship reported that he had observed such use of a bottom seine by Japanese fishing vessels officially engaged in pollack fishing in Russian waters in the Bering Sea.

Unauthorised production by crew members

Some crew members are involved in the unauthorised production of high-priced products. This is a low-volume activity usually involving salmon roe and crabs. Most vessels have home made "crab-stoves" for cooking crabs. Usually these consist of a standard 40-litre metal milk churn with two to three heating elements fitted inside. The meat from cooked crabs is frozen in cold storage and then hidden amongst the products in the hold. Some fishing vessels that lack freezing facilities have unregistered canning machines to seal crabs or salmon roe in tins. There are crew members who buy the crab and roe on virtually all the processing ships that take on raw seafood from small fishing vessels. These buyers have permanent trading partners in foreign ports or on foreign carrier vessels. Payment is made in cash (USD) on the spot, as for other types of illegal trade in fish products. In the summer of 1999, the average income from this activity per crew member was estimated at about RUR2 000 (USD 100) per day. If their earnings were to drop to RUR1 000 (USD 50) per day, this illegal activity would no longer be considered profitable enough to warrant the risk (TRAFFIC survey data).

Credit: Alexey Vaisman, TRAFFIC Europe-Russia

Home-made electrically heated pot used to boil illegally caught king crabs

In some cases, the income from personal production constitutes the crew's wages: companies with catch quotas have frequently issued contracts to fishing vessels whereby expenses such as repairs, fuel and water are paid, but not wages. The income of crew members is generated from what they catch and sell themselves, over and above the quotas obtained (TRAFFIC survey data).

Pollution at sea

This activity involves unauthorised dumping of production and household waste. Under the current rules governing fillet production and roe collection, all discards must be chopped up to the consistency of mulch before being dumped over the side. In practice, industrial fishermen do their utmost to simplify this lengthy procedure and, if possible, do without it altogether (TRAFFIC survey data). As noted in an earlier chapter, production of Alaska Pollack products newly in demand which use only part of the whole fish have resulted in increased dumping of discards into the sea, by Russian, American, Japanese, Chinese, Polish, and South Korean vessels sailing in the western Bering Sea (TRAFFIC survey data).

Fishing without a licence

The captain of a vessel is responsible for acquiring a licence and is therefore considered the offender if this is missing. However, this form of law breaking accounts for only one per cent of all fishing violations. Committing an offence under almost any of the categories described so far is greatly facilitated if the ship's captain has an official permit to fish, if only for one tonne (*Kamchatrybvod* unpublished).

Factors contributing to illegal activities

Besides the initial motivation to supplement or earn income, there are several factors conducive to illegal fishing-related activites in the Russian part of the Bering Sea, as outlined below.

Inappropriate legislation

Out-of-date legislation

Russian fisheries legislation is not tackling the issues which are placing marine resources in the Bering Sea under threat. In the new market-led context, it has proved ineffective where inappropriate anachronistic laws from the Soviet system are still in place. For example, according to the law, police or other inspectors only have the right to institute criminal procedures against those committing fisheries offences at the request of a ship owner or fishing company staff, who are in many cases the very perpetrators of the crime. As a consequence of this (as well as other factors) the very large majority of infractions that are discovered by the police are not followed up.

Gaps in legislation

To add to the problem of outdated legislation is the fact that some aspects of fishery regulation in the Russian part of the Bering Sea are not addressed legislatively at all. For example, according to current regulations, marine bioresources harvested in the Russian EEZ and on the continental shelf of the Russian Federation, but outside its territorial waters, are not subject to normal Customs clearance procedures, provided they are caught and sold without entering Russian territorial waters (see **National legislation**). The discrepancy between laws governing Customs clearance for fish and other seafood caught in Russian territorial waters and laws governing catches of the same beyond the territorial limit, (but within the EEZ) has made development of illegal fishing and subsequent uncontrolled sale of catch easy. Further examples include the need for legislation to regulate use of some fishing gears, such as bottom-nets and drift-nets, the need for a tighter legal rein to be applied to the establishment of joint ventures and to the process for transfer of nationality to fishing vessels.

High import and export duties

Presidents of major fishery associations of the Russian Far East and managing directors of several leading fishing and processing companies of the region claim that export duties are rising. They feel that they are lacking in State support by contrast with their counterparts in other countries, where tariff and tax policies are aimed at supporting the national producer, as for example in Japan (Anon., 1999i). This perception presumably only serves to encourage avoidance of duties through methods such as trans-shipment of catches beyond Russian territorial waters.

High import duties are also conducive to law-breaking. Duties payable for ships imported into the Russian Federation are extremely high and must be paid immediately. This makes it unprofitable to import vessels made abroad, yet Russian fishing enterprises need modern, well-equipped craft built outside the Russian Federation. A very simple solution to this has been found in purchasing vessels which are never brought within Russian territorial waters, or at least not to a Russian port. As a result, a large number of foreign-made vessels fish under a Russian flag in the Russian EEZ but never put in to Russian ports, nor even into Russian territorial waters. This encouragement of fishing beyond territorial waters is likely to increase the number of cases of illegal fishing and sale beyond the control of Russian law.

Low penalties

Penalties associated with infractions appear to be too lenient. For example, one of the highest penalties for a fisheries offence is the imposition of a fine on the captain of a vessel for theft of fishing equipment. This fine is usually 200-500 times the "minimum wage" (where this is in fact not a wage as such but an official figure used as a unit for calculating fines: in late 1999, this figure was the equivalent of about USD4-4.5). In any case, offenders generally pay little or no attention to fines because the profits from illegal activities usually exceed the fines many times over.

OBSERVERS: The job of observer (inspectors permanently based on board foreign vessels fishing in Russian waters of the Bering Sea) can be an attractive one; the wages can be high. Expenses incurred while on board, as well as the salary, are paid by the ship owner or the firm leasing the vessels. In simple terms, inspectors are paid by the firms they are supposed to be monitoring. Salaries depend greatly on the country and the particular fishing company to which the vessel belongs. Usually, the on-board inspectors try their best not to reveal their earnings, but reportedly Japanese ship owners pay up to USD120 per day. The South Korean ship owners reportedly pay from USD80 to USD100 per day, while Norwegian and Taiwanese owners pay slightly less. Polish ship owners are reported to pay approximately USD40-50 per day, while the Chinese have the "worst" reputation in this regard, paying only USD20-25 per day. In addition, the on-board inspectors have free access to food and alcohol. There are rumours that some firms especially interested in having good relations with the inspectors tend to satisfy their "fleshy desires" during the trip (TRAFFIC survey data).

Weaknesses in the enforcement system

A number of factors were found to contribute to the poor performance of Russian enforcement bodies, as outlined.

Poor co-ordination between enforcement agencies

One of the main weaknesses in the system is the poor co-ordination and institutional rivalry among the bodies which share reponsibility for enforcement. In the western Bering Sea, *Kamchatrybvod*, the Special Marine Inspectorate and the border control units, which have responsibility for protecting not only Russian territorial waters but also marine resources in the EEZ, do not agree on each other's role and do not co-ordinate activities. On the contrary, difficult tasks, such as combatting poaching at sea, are sometimes avoided by one of these agencies by referring the responsibility to one of the others (TRAFFIC survey data). An example of such an incident is provided by the case in April 1999 of an inspection vessel which contacted *Kamchatrybvod* after detecting unauthorized crab fishing. Rather than tackle the problem, *Kamchatrybvod* passed it on to the Federal Border Service, which reported that it was not in a position to respond to the call for assistance and referred the matter back to *Kamchatrybvod*.

Corruption among enforcement staff

Within the enforcement agencies themselves, there is evidence of corruption among the staff. During survey work for this report, such corruption was reported to be common. According to one Russian Government source, "tolerance for poaching, inertia and at times direct protection of poachers by inspectors of the fisheries agencies have become a....problem. Bribery of government inspectors and their use as a cover for illegal fishing by groups of vessels or individual companies has become a widespread phenomenon" (Regional Directorate of Interior, unpublished, July 1999).

Corruption is undoubtedly an obstacle to enforcement among observers on board most foreign fishing and factory ships that operate in the Russian part of the Bering Sea. The system for appointing and remunerating observers is conducive to corrupt practice. Observers may or may not be permanent staff members of fish protection agencies. Those in the latter category usually work at the scientific research institutes of the State Fisheries Committee or are sometimes simply family members or friends of fisheries law enforcement agents (TRAFFIC survey data). There are no pre-determined earnings for observers and the extent of financial reward varies, partly according to an observer's level of activity on board, reportedly. An assignment on a "good" ship is considered by an observer to be a reward, often expected to be recompensed in turn, with a gift or money from the observer to his superior. Fishing companies with a constant presence in the Russian EEZ have reportedly established agreements with observers and their bosses about the way the system should operate and those observers who try to prevent or disclose violations are excluded from the on-board observers "family", or at least assigned to Chinese vessels (see box on observers). The discovery in autumn 2000 of a Japanese vessel inside the Commander Islands prohibited zone with an observer on board, is testimony to the token nature common to this post and even to its detrimental potential in so far as it could provide a means to "legalize" catches with a signature from a corrupt observer.

Poor state of government equipment

For those enforcement officials who do attempt to fulfil their tasks and duties, competing with the well-financed and well-equipped commercial vessels is exceedingly difficult given the equipment with which they must operate. Offices do not have computerized systems and inspection ships are outdated and too slow to catch offenders equipped with modern high-speed vessels. In April 1999, when the inspection vessel *Dalliya* pursued unauthorized crab fishers in waters off Kamchatka, its engine burnt out during the chase. Equipment problems continued to confound enforcement when the Federal Border Service was called upon to assist. They had not a single marine protection vessel in the area at the time, the nearest being off the south Kurils, from where the voyage would take at least two days. Aircraft from the Service could not take off because of adverse weather conditions (S. Vakhrin, *Kamchatrybvod*, pers. comm,. 1999).

According to *Kamchatrybvod* staff, fishermen monitor the movements of inspection vessels by radio and openly exchange the information (Anon., 1998f). The best hope for inspectors is often to approach suspected poachers clandestinely under the cover of darkness or mist.

Credit: Alexey Vaisman, TRAFFIC Europe-Russia

The inspection ship *Ingeneer Martynov* follows in the wake of arrested ship

Low pay for enforcement staff

Government fisheries inspectors receive extremely low wages. The average wage of an inspector is just RUR800 a month, an amount equivalent in value to a month's basic groceries for one person. Such meagre wages could increase temptation to obtain extra income from illegal activity (TRAFFIC survey data).

Lapse in control during re-organization of the system

In 1997, the fisheries enforcement system was re-organized in accordance with presidential *Decree No.950*. Several key functions, some permanent staff and materials and equipment from the State Fisheries Committee of the Russian Federation were transferred to the Federal Border Service. The lengthy reorganization process in 1997 and 1998 meant that fisheries enforcement, especially in 1998, was much reduced and equipment and experienced staff were lost.

Organized crime

The operations involving illegal fishing appear to have taken on an increasingly organized character from year to year, although quantifying the level of "professional poaching" is difficult. Stakes can be high in illegal fishing activities and many of those involved were hesitant to speak to investigators on the subject. The following examples illustrate the organized nature of the illegal fishery.

- During this study, the existence of a type of document referred to as a "**provisional instruction**" came to light. Captains of vessels are instructed by criminal organizations to "read through, memorise and destroy" such information prior to embarking on an expedition. To aid secrecy, references to certain fishing areas are in code, recommendations for encoding the co-ordinates of locations are made and instructions provided on keeping fictitious and actual records in the ship's log and the fishing and engine logs. In an excerpt from one document issuing guidance on illegal practice, captains were instructed to "maintain the ship's log and the fishing and engine logs strictly as agreed, complete fair versions, including a lag to allow for the time necessary for the transition from the fictitious to the actual area of operation. Each day, captains must give directions to the person in charge of production on how the documents on product yield have to be filled out."

- Another reflection of the organized nature of the "professional poachers" is the way in which small **groups of vessels divide tasks** and responsibilities amongst themselves so as to evade the law more effectively. One vessel takes a co-ordinating role, using navigational equipment to check the whereabouts of the patrol vessels in the fishing area, while directing the activities of the other vessels so that they can disperse as quickly as possible at the approach of an inspector. During investigations for this report, intensive radio communication between the ships of such a poaching group was witnessed, during which the co-ordinates and external appearance of the inspection vessel that was approaching were broadcast. The poaching vessels had hidden their names and registration numbers behind panels that had been made in advance to render identification more difficult. One captain of a factory ship apprehended by the Special Marine Inspection Service admitted in writing that he had been the organizer of illegal fishing activities in prohibited areas for a joint stock holding company. He admitted co-ordinating the process and communicating messages about the location of patrol vessels. The captains of two other vessels involved later corroborated the account given by the captain of the factory ship.

CONCLUSION

Before drawing conclusions from this investigation, it is important to reiterate that interpretation of the information is complicated by the fact that it is known that much is missing and some potentially unsound. Official catch and export data, for example, are unreliable and definitely gross underestimates of actual catch and export volumes. Even were these to be known, it is not possible to assess how resources might be affected, since stock data are either unavailable or insufficient and out of date.

Such information as exists on stocks in the western Bering Sea indicates that the biomass of Alaska Pollack has almost halved from 1991 to 1996; stocks of Pacific Cod have remained steady in number between 1978 and 1998, and herring stocks are increasing. For Saffron Cod, rockfishes, halibuts and flounders, king crabs, shrimps and squid little or no information is available to determine whether stocks are currently increasing or decreasing in the Russian part of the Bering Sea.

Catch data for the western Bering Sea from 1980 to 1993 indicate that recorded catches overall were relatively steady, averaging about one million tonnes annually, although catches for some species - for example, cods, herring, crabs and prawns - sometimes showed wide variations between years. The size of Russian catches from the western Bering Sea after 1993 are not known as data are unavailable, but statistics for Russian catches in the wider Pacific Northwest area show that overall declines were apparent over the years 1984-98 for Alaska Pollack, Pacific Cod and Pacific Ocean Perch. The same data source (FAO) reveals that Russian catches of herring in the last years of the 1990s were at least double the volume of such catches for several years in the 1980s. Russian *Pandalus* shrimp catches displayed a definite upward trend between 1988 and 1998. Comparison of catch levels in the Pacific Northwest and the Pacific Northeast reveals that the volumes for certain fisheries products can vary widely between the two areas. Recorded catches of Alaska Pollack, Pacific Herring, Pacific Ocean Perch and king crab, for example, have often been several times greater in one of these areas than in the other, during the period 1984-1999. This highlights the importance of assessing conservation of resources for the Bering Sea as a whole, where fish and other organisms move across the entire ecoregion, and where remedial action to relieve pressure on stocks within national boundaries will achieve only limited success.

The vagueness of available knowledge of fisheries stocks and the related impact of harvests from the western Bering Sea is in itself a conclusion of this study. Based on this fact, it may further be concluded that the management procedures for fisheries in the western Bering Sea are set on obscure foundations. The overall quota for marine resources examined in this report has risen by 64% between 1996 and 1999 (from 581 630 t to 957 896 t). Although much of this extra amount consisted of greater allowances for herring and shrimp, which can reportedly withstand increased fishing in the western Bering Sea, there is a lack of transparency to quota setting. As such, the basis for an increase in the western Bering Sea quota for Alaska Pollack, which was apparently set against a background of decline in stocks of the species, is unclear. This combination of circumstances is made worse by the fact that the legal quota is far exceeded by illegal fishing.

The potential seriousness of this situation is underlined by the ecological and economic importance of the region. The Bering Sea is estimated to have contributed about half of the Russian marine harvest during the 1990s and its productivity is key to the economic and social stability of the Russian Far East. It is also judged to be one of the most important regions for conserving diversity of life on earth and yet minimum effort at government level is directed to safeguarding the Sea's resources at present. Stocks of Alaska Pollack, the fish which has dominated the catch from the western Bering Sea throughout the 1980s and into the 1990s, are not being managed responsibly, despite having underpinned the fisheries sector and in turn the economy of Kamchatka. Together with cod and herring, this species accounted for over 90% of the western Bering Sea quota in the second half of the 1990s. Pollack, on which the region's fishery depends, has diminished in Russian catches from the Pacific Northwest, 1984-98, and the same is true of cod. King crabs, the most lucrative catch in terms of export value per kilogramme, may be in decline in the western Bering Sea according to scientists, and as suggested by much lower quotas for the western Bering Sea in 1998 and 1999 than in the previous two years.

Against this background of uncertain stock levels and uncontrolled catches, exports of marine resources from Kamchatka have been rising. The total value of the marine products exported from Kamchatka has risen sharply from 1995 to March 1999. In 1995, exports were valued at over USD96 million and by 1998 this figure had grown to more than USD165 million. This increase in the total value of exports corresponded to an increase in volume of exports, on the whole - the overall volume of exports rose from 37 414 t in 1995 to 201 388 t in 1998. The surge in exports can be attributed in part to changes in the socio-economic status of the fisheries sector and, in particular, to the fisheries sector becoming export-oriented in an effort to earn hard currency during an era of economic collapse. The main recipients of this flow of marine resources from the Russian Bering Sea, in terms of volume, were South Korea, Japan and China. While the volume and overall value of exports have risen, however, data from Kamchatka Customs indicate that the export price per kilogramme of most species has declined steadily, with the exception of Alaska Pollack, squid, halibut and Atka Mackerel. This may be partly because of the dismantling of the fisheries processing industry, leading to the export of lower-value raw products instead of processed commodities Additionally, the unit cost and price of exports fell as exporters fled Russian territorial waters during the 1990s to circumvent taxes, but these transactions carried out beyond the control of authorities will not be recorded in official Customs statistics.

While exports of marine products caught in the Russian part of the Bering Sea have risen in recent years, the corresponding contribution to the budget has not. Since 1995, Kamchatka's fishing sector has run at a deficit, suffering from a sharp drop in production potential. Beset by the problems of deteriorating equipment, higher operating costs, and the general economic instability following the change in political regime at the outset of the 1990s, many in the fisheries business of the western Bering Sea have resorted to illegal methods of earning income. The existence of illicit activities, including catching, processing and storage of fish in excess of quotas, is significant at all levels, from sailors on small isolated fishing vessels to organized fleets of large industrial fishing boats. It is apparent that a large part of the catch is spirited away without being entered in the ledgers, a process assisted by the fact that Customs regulations differ depending on whether exports are from inside or outside territorial waters (within 12 nautical miles from shore). The importance of this discrepancy in regulation is stressed by the fact that the volume of catch made in the Russian EEZ from beyond territorial waters was four times greater in 1995 than the volume that cleared Customs. For Alaska Pollack in the Kamchatka region, the volume of illegal catch is estimated to range between 15% and 50% of the volume of the legal quota. The entering of false information on official documents, which in turn facilitates other illegal activities, such as storing and selling surplus catch, was perceived to be a widely used device. Other acts against the law, such as fishing in areas officially out of bounds, are sometimes facilitated by the presence of the very observers installed on vessels to police the fishery. This points to the highly organized nature of much illegal activity in Russian fisheries of the area: most criminal acts in recent years appear to have been committed by a mere handful of companies.

Estimates for the value of the overall illegal catch in the Russian Federation range from one to five billion US dollars each year. If the value of such losses from the western Bering Sea are in proportion to the region's share of the overall national legal catch, then they would amount to USD0.5-2.5 billion annually.

The regulatory framework to address these problems is unsatisfactory and riddled with gaps, anachronisms and ill-advised policies. Most fisheries management issues in the western part of the Bering Sea and its adjacent seas are, however, not a matter of adopting additional legislation, but rather of appropriate implementation of existing regulations. The level of surveillance of fishing in the Russian EEZ, for example, is not equal to the scale of the problem. Government agencies in charge of protecting the country's marine resources do not co-ordinate their activities, nor are they provided with the necessary resources. On-board government inspectors are remunerated by the firms they inspect and their duties are degenerating into pleasant and profitable ways of spending time. Under such conditions, an inspector's main duty becomes to ignore violations.

To sum up, the current lack of shared data on the status of marine resources of great biological and economical importance is a major obstacle to the management of fish stocks in the western Bering Sea. There is no sound platform from which to judge the effects of the intense fishing in this diverse marine environment, but instead of taking a precautionary approach through the implementation of more conservative quotas and stronger enforcement regulations, the Russian

Government is struggling to control fishing activities within its economic zone against a host of obstacles, including organized crime. The conservation of marine ecosystems is a trans-border issue, however, since responsible stewardship of the Bering Sea should take into account the needs of fishers from several countries as well as actions necessary to ensure healthy and diverse marine life. What is happening in the present, by contrast, is unchecked and unconstrained extraction from an outstanding environment for short-term, financial profit.

RECOMMENDATIONS

As this report addresses industrial fisheries, these recommendations should, for the most part, be considered in relation to the large fleets of the industrial fishery of the Russian part of the Bering Sea.

Fisheries management in the Russian Federation

Fisheries information

- Surveys of fish stocks, with species-specific information, should continue to be conducted and enhanced. These should ideally be full-scale surveys on an annual basis, but it is recognized that this is an expensive task for any country. Therefore, if funds do not permit full-scale surveys, research should be focused on priority areas, including collecting baseline information for species about which little is currently known. Monetary grants should be allocated to specified research priorities, co-ordinated as part of a strategic research plan spanning a defined time period, for example five years. In the absence of necessary funds, the presence of a scientist (for example, an experienced staff member from the Pacific Research Institute of Fisheries and Oceanography (TINRO)) should be required on industrial fishing vessels during their expeditions to collect biological data and record information on Catch per Unit Effort (CPUE). The geographical deployment of these scientists should be carefully planned in order for their reports to cover the most representative fishing grounds and provide information on the most exploited (most in demand or most depleted) fisheries resources.

- Scientific results of surveys of fish stocks should be published in regularly issued proceedings under the auspices of the proposed bilateral Russian-US fisheries management agreement (see below).

- Closer management of funds raised for stock research (for example, from sale of "controlled catches") should be initiated through the establishment of a separate budget line for this purpose. Additionally, a marine resource research programme at federal level should be developed.

- Improved procedures and, where needed, administrative structures should be created to allow for the most accurate, reliable and up-to-date catch data for species targeted by Russian industrial fishing vessels to be reported to the UN Food and Agriculture

Organization (FAO). Although the Russian Federation is not a member of FAO, effective reporting could be facilitated by reverting to the relationship of close collaboration which existed between FAO and the Soviet Union. In this way, the data will be available to neighbouring countries and to international institutions, that need them to take adequate management measures for shared fisheries stocks (for example, the setting of appropriate catch quotas).

Management of stocks

- Consistent with its undertakings under the UN fish stocks agreement, the Russian Government should adopt a precautionary approach to the management of industrial fisheries in the Bering Sea, particularly given the high level of uncertainty regarding the status of some stocks. This would include the establishment of stock-specific precautionary reference points and harvest levels that take into account the estimated level of illegal catch.

- The criteria for quota allocations should be made transparent to stakeholders. This would have the benefit of being an overt system, reducing perceptions of unfair allocations and disclosing biological considerations taken into account. A list of quota recipients could be made widely accessible *via* an electronic database. As a further development of the database, a field for recording catches against quotas could be included. An appropriate agency should be appointed by the Government of the Russian Federation to undertake the task of developing this database.

Protected areas

- Key habitats, especially those that nurture specific biological processes in the western Bering Sea, should be identified and fishing prohibited in these locations during critical seasons. These would be likely to include wintering, spawning and nursery grounds of heavily exploited and/or apparently depleted fishery stocks.

Fishing gears

- Japan's continuing use of drift-nets in Russian waters should be addressed, acknowledging the problems caused by large-scale drift-nets that led to the resolution of the UN General Assembly on large-scale pelagic drift-net fishing and its impact on the living marine resources of the world's oceans and seas.

- Regulations relating to fishing gears should be extended to prohibit all non-selective and destructive gear, without exclusions and reservations.

Strengthening of governance specifically in the Russian EEZ of the Bering Sea

- A revision of Russian Customs regulations should be initiated, with the aim of adapting them to present-day marine fisheries, including Russian exports of marine products. This revision could be initiated by the State Customs Committee and should harmonize laws for all fisheries exports, whether from inside or outside the territorial limit of 12 nautical miles from shore.

- Legislation applying to the establishment of joint ventures and to the transfer of nationality of vessels should be tightened where necessary, with the aim of preventing opportunistic choices of nationality and uses of fishing quotas.

- Revision of the Russian shipping register should be undertaken, to prevent foreign-purchased vessels fishing under the Russian flag without ever entering Russian territory, thereby avoiding the closer regulation and also the costs associated with coming into port.

Social and community considerations

- People living adjacent to the Bering Sea should be involved in decision-making affecting the resources on which they rely. Unless they have a stake in developing conservation plans for the region, progress to safeguard the Bering Sea and its life forms may be undermined. Furthermore, communities which have populated the coastal areas of the Bering Sea for decades or longer are in a position to offer valuable information on changes witnessed in the Sea and in the status of its resources.

- Regional fisheries should balance the interests of industrial fishing fleets against the needs of the coastal fishing and processing industries. Better support of these local industries will provide work for coastal communities which could prove an attractive alternative to poaching from the sea. Support for such coastal-based industries should be provided in the form of tax exemptions and reserved catch quotas.

Finances

- A system should be developed for channelling a proportion of fines for fisheries infractions back into enforcement of fisheries-related law. This system could facilitate the provision of funding for some of the changes to fisheries control recommended.

- The Government of the Russian Federation should consider using fisheries-related income from sources other than fines, for example, money from quota sales, to finance some of the reforms necessary in the fishing industry of the Bering Sea. Co-ordination of the observer network and training of observers, for example, could be funded in this way.

Enforcement action in the Russian Federation

Co-ordination of agencies

- Clear and "non-overlapping" terms of reference should be formulated for each agency involved in enforcement of Russian fisheries and trade laws. Specific reporting obligations between each agency, as well as to federal authorities, should be included in the terms of reference, with precise time-frames specified.

Improvement of the system of observers

- An official network of observers should be created and co-ordinated, with new operating conditions to reduce the opportunites for corruption inherent in the current system of observers. For example,

  - Placement of observers should be on a rotational basis throughout the various fleets, with the aim of reducing the chance for corruption to be fostered within a long-term working relationship between one observer and one company and/or crew.

  - The salaries of observers should be paid only *via* an official administrative body. There should be no direct payments by vessel owners to observers.

- Training programmes for observers should be set up. Minimum criteria and standards should be established that must be met before individuals can be considered as observers. Specifically, training should be undertaken in methods commonly used to falsify documents and other means of avoiding legal restrictions.

- Observer coverage should be expanded to include domestic vessels, in recognition of the fact that Russian vessels appear to be implicated in the majority of offences and also that the nationality of vessels is easily transferable in some cases. Apart from boosting enforcement potential, this would also provide a means to collect comprehensive catch data for eventual comparison against quotas.

- The possibility should be considered of observers acting not only as fisheries inspectors, but also as Customs inspectors, where applicable.

Improvements in equipment to aid enforcement

- Existing satellite monitoring systems should incorporate modern, remote-access technology, such as that developed by technical staff within *Kamchatrybvod*. Additional aids to monitoring, such as mechanisms for gauging fuel consumption and use of a video monitoring system should be considered. The cross-referencing by enforcement personnel of any information on vessel positions with information stored in the proposed database of quota allocations should be made possible.

- Old and inadequate equipment used by the enforcement agencies should be replaced with modern equivalents.

Adjustment of financial incentives and disincentives

- Penalties for fisheries offences should be increased, by adapting them to take account of the fact that fishing in the Russian part of the Bering Sea is now a commercial industry and that profits from illegal fishing can be high.

- The feasibility of a transparent and direct bonus system for enforcement staff, based on rewards for increased seizures, should be assessed. The remuneration system has to make bribe-taking less profitable than honest service.

International co-operation

- Interaction between Customs agencies of countries trading in Bering Sea marine resources should be improved. Information on exports and imports of these resources between trading partners should be shared in order to identify discrepancies in trade volumes and so indicate illegal trade. Additionally, Customs agents in importing countries could be requested to ask captains of Russian vessels exporting catches direct from the Russian EEZ to show a copy of the official Russian documentation authorizing these. The appropriate Russian authorities could then be informed of any failure to produce such a document and take action accordingly.

- As already implemented by numerous countries, and in order to permit exchange of information on a permanent basis regarding volumes of exports and imports of particular marine species and their various products, all nations involved in trade in Bering Sea fishery products should apply the most precise category code available of the Harmonized System (Harmonized Commodity Description and Coding System). For example, at species level, the code for frozen fillets of Alaska Pollack *Theragra chalcogramma* is 0304 20 85. Where necessary, new codes unique to species should be adopted. The State agency in charge of compiling statistical data on trade and other economic information at national or federal level should be responsible for preparing an annual report in a form that is accessible internationally. An official reporting procedure should be established so that agencies at regional or provincial level send Customs trade data of the previous year to the designated State agency within a strict time-frame, for preparation in the annual report. Advice can be provided on existing systems by WCO (World Customs Organizations).

- The Parties of the *Convention on the Conservation and Management of Pollack Resources in the Central Bering Sea* should consider the implications of closure of the Donut Hole to Alaska Pollack fishing, in particular with regard to increased pressure that is placed on the marine resources in the western Bering Sea.

- As unilateral decisions often have an immediate and negative impact on the control of fisheries in a neighbouring State, the importance of bilateral US-Russian decision-making should be emphasized. The Russian Federation and the USA should initiate negotiations to develop a permanent bilateral agreement for a joint approach to fisheries management, so that proposals for new national regulations on Bering Sea fisheries, for both the Russian Federation and the USA, will be planned in a co-operative way. These should include, for example, new limits on the fishing capacity of industrial fishing vessels, prohibition of certain types of fishing gear or fishing prohibitions in certain areas, during certain months.

Awareness

- A conference bringing together industry, regulatory agencies and environmental non-governmental organizations should be convened, to discuss problems linked to the conservation, management and sustainable use of marine resources in the western Bering Sea.

- Information on the levels of threat to fish stocks should be made available to interested parties. This could attract the involvement of non-governmental groups (including industry), such as direct funding or lobbying for increased resources to be directed towards certain stocks.

- Economic incentives for the promotion of sustainable fisheries as a marketing tool should be developed. Discussions on such possibilities should be held at an international level with organizations such as the Marine Stewardship Council, towards development of schemes for certification (and "eco-labelling") of sustainably harvested fish stocks.

REFERENCES

Anon. (1985). King Crab: *Paralithodes camschatica, P. platypus* and *aquispina*. Pp. 387-399. In: Anon. (1985). *Alaska Habitat Management Guide, Southcentral Region, Volume 1: Life Histories and Habitat Requirements of Fish and Wildlife*. State of Alaska Department of Fish and Game, Division of Habitat, Juneau, Alaska.

Anon.,. (1994a). Review of the state of world marine fishery resources. *FAO Fisheries Technical Paper 335*. FAO Fisheries Department, FAO, Rome.

Anon. (1994b). Tanner Crabs. ADF&G Wildlife Notebook Series. Alaska Department of Fish and Game. http://www.state.ak.us/adfg/notebook/shellfsh/tanner.htm. Viewed 31 October 2000.

Anon. (1995). The world market for crab. *FAO/GLOBEFISH Research Programme* 37. Rome, FAO. Pp. 59.

Anon. (1996a). *The Bering Sea Ecosystem*. The National Academy of Sciences, National Academy Press, Washington, DC.

Anon. (1996b). *Kamchatka Fisherman*. 26/7/96.

Anon. (1997). *Biodiversity Conservation in Russia*. Project GEF "Biodiversity Conservation". The State Committee of the Russian Federation for Environmental Protection, Moscow.

Anon. (1998a). *The Times Atlas of the World*. Comprehensive ed. 9th ed. Times Books, London.

Anon. (1998b). Russia News: Revision of quota allocation underway. *Eastfish Magazine* 3/98:10.

Anon. (1998c). International News: Russia could double fish production. *Eastfish Magazine* 5-6/98:6.

Anon. (1998d). Guide to Rockfishes (Scorpaenidae) of the Genera *Sebastes, Sebastolobus*, and *Adelosebastes* of the Northeast Pacific Ocean. *NOAA Tech. Memo. NMFS-AFSC-95*. 1998. Adapted from Orr, J.W., Brown, M.A., and D.C. Baker, (1998). *Guide to Rockfishes (Scorpaenidae) of the Genera* Sebastes, Sebastolobus, *and* Adelosebastes *of the Northeast Pacific Ocean*. http://www.afsc.noaa.gov/techmemos/nmfs-asfc-95.htm.

Anon. (1998e). *Kamchatka Fisherman*. 22/1/98.

Anon. (1998f). Licensing as a way of making ghosts materialize. *Kamchatka Fisherman* 12. 19/3/98.

Anon. (1998g). Russia News: Chinese vessel detained with illegal pollack catch. *Eastfish Magazine* 2/98:11

Anon. (1999). *The State of World Fisheries and Aquaculture 1998*. FAO, Rome, Italy. In: Groombridge, B. and Jenkins, M.D. (2000).

Anon. (1999a). Stable catches mask stock declines in Russia. *WorldFish Report* 94:FS/2, 17 June 1999.

Anon. (1999b). Report of the Kamchatka Region Statistics Committee. Unpublished report.

Anon. (1999c). Russian fish prices pushed up by tax. *WorldFish Report* SP/2, August 11.

Anon. (1999d). *The 1999/2000 Directory: Russian Far East Fishing Companies*. Pacific Rim Fisheries Program, Institute of the North, Alaska Pacific University, Anchorage, Alaska.

Anon. (1999e). Russia News: Moscow publishes 1998 catch and production results. *Eurofish Magazine* 3/99:14-15.

Anon. (1999f). *Surimi* - new decanter removes bottleneck of de-watering. *Eurofish Magazine* 4/99:38.

Anon. (1999g). *Annual Commodity Update: Crab.* FAO Eastfish, Copenhagen, Denmark.

Anon. (1999h). Russian scientists highlight squid fishery potential. *WorldFish Report* 92:SP/1, June 3, 1999.

Anon. (1999i). Russian fishermen go for seats in Parliament. *WorldFish Report* FS/4, August 25.

Anon. (2000a). *International Agreements Concerning Living Marine Resources of Interest to NOAA Fisheries.* International Fisheries Division, Office of Sustainable Fisheries, National Marine Fisheries Service, NOAA, Silver Spring, Maryland.

Anon. (2000b). Russian Far East: Primorye protests fish quota cut. *Novosti, The Vladivostok News*, 3 March, in *Pacific Rim Fisheries Update* 9(43):4.

Anon. (2000c). Fishing Industry News: Sales of quotas to foreigners may be discontinued. *Russian Fish Report* 4(43):3.

Anon. (2000d)). Russian Far East: RFE King crab quotas cut. *Sakhalin Fisherman* 12/2000, 24 March, in *Pacific Rim Fisheries Update* 9(43):3.

Anon. (2000e)). Russia News: Overview of overall Russian catches in 1999. *Eurofish Magazine* 3/2000:16.

Anon. (2000f). Species Summary for *Theragra chalcogramma* Alaska Pollack. Updated 12/10/2000. http://www.fishbase.org/SpeciesSummary.cfm?ID=318& genusname=Theragra&speciesname=chalcogramma.

Anon. (2000g). Fishing Industry News: Kamchatka scientists call for reduction of pollack catches. *Russian Fish Report* 4(43):5.

Anon. (2000h)). Russian Far East: Committee intervenes in pollack catch. *Primorye Fisherman* 9/2000, 3 March, in *Pacific Rim Fisheries Update* 9(43):2.

Anon. (2000i)). FIS - Market Reports: Bottomfish Market Reports. 30 September 2000. http://www.fis.com/fis/reports/report.asp?l=e&species=16&date=.

Anon. (2000j)). Species Summary for *Gadus macrocephalus* Pacific Cod. Updated 12/10/2000. http://www.fishbase.org/SpeciesSummary.cfm?ID=308&genusname=Gadus&speciesname=macrocephalus.

Anon. (2000k). Species Summary for *Eleginus gracilis* Saffron Cod. Updated 12/10/2000. http://www.fishbase.org/SpeciesSummary.cfm?ID=315&genusname=Eleginus&speciesname=gracilis.

Anon. (2000l). Species Summary for *Clupea pallasii* Pacific Herring. Updated 12/10/2000. http://www.fishbase.org/SpeciesSummary.cfm?ID=1520&genusname=Clupea&speciesname=pallasii.

Anon., (2000m). Market Monitor: Scientists publish TAC forecast for the year 2000. *Russian Fish Report* 5(44): 19-31.

Anon. (2000n). FIS - Market Reports: Squid/Octopus Market Reports. 20 September 2000. http://www.fis.com/fis/reports/report.asp?1=e&specie=256&date=.

Anon. (2000o). (FIS - Market Reports: Squid/Octopus Market Reports. 19 October 2000). (http://www.fis.com/fis/reports/report.asp?1=e&specie=256&date=.

Anon. (2000p). Territorial dispute overshadows Russian talks with Japan. *Russian Fish Report* 9(48):1.

Anon. (2000q). Poachers fined $2.5 million through control in Japan. *Russian Fish Report* 8(47):3.

Anon. (2000r). Japanese boats to pay RUR 40 million in damages. *Russian Fish Report* 8(47):4.

Anon. (2000s). Nakhodka court confiscates Japanese boat. *Russian Fish Report* 9(48):6.

Anon. (2001). *Fishing News International* 40(7): 1. Heighway, a division of AGRA Europe (London) Ltd - Informa Group Plc.

Bakilin, V. and Ignatenko, O. (1996). Crab fever. *Severnaya Pacifica* 1: 39-45.

Balykin, P.A. (1990). *Biology and the State of Mintai Resources in the Western Bering Sea.* Abstract of a thesis for a PhD in biology. Pp. 23.

Balykin, P.A. (1986). Mintai fish productivity in the western Bering Sea. *Ichthyology Issues* 26(1): 164-168.

Banks, D., Williams, M., Pearce, J., Springer, A., Hagenstein, R., and Olsen, D. (eds.). (1999). *Ecoregion-Based Conservation in the Bering Sea: Identifying Important Areas for Biodiversity Conservation.* World Wildlife Fund, Washington, DC, and The Nature Conservancy of Alaska, Anchorage, Alaska.

Bizikov, V. A. (1996). The Komander squid in the Bering Sea: its role in ecosystems and biocenological ties. Pp. 6-8 in *The History and Perspectives of Fisheries and Research Objectives: Commercial Aspects of Biology of Komander Squid and Slope Fish in the Western Part of the Bering Sea.* VNIRO, Moscow.

Blau, S.F. (1997). Alaska King Crabs. ADF&G Wildlife Notebook Series. Alaska Department of Fish and Game. http://www.state.ak.us/adfg/notebook/shellfsh/king crab.htm. Viewed 31 October 2000.

Cline, D. and Williams, M. (eds) (2000). *The Bering Sea Ecoregion. A Call to Action in Marine Conservation.* WWF-US, Washington D.C. and Beringia Conservation Program, Anchorage, USA.

Day, C. (2000). *Berryteuthis magister magister*. Updated 15 September 2000. http://www.cephbase.dal.ca/speciesc.cfm?CephlD=334.

Fossbakk, T. (2000a). Poaching and lack of patrols damaging Russian fisheries. *FIS Hot News*, 7 July 2000. http://fis.com/fis/hotnews/.

Fossbakk, T. (2000b). Japanese and Russian caught fishing illegally. *FIS Hotnews*, 6 July 2000. http://fis.com/fis/hotnews/.

Groombridge, B. and Jenkins, M.D. (2000). *Global Biodiversity. Earth's Living Resources in the 21st Century*. World Conservation Press, Cambridge, UK.

Hodgson, B. (1992). Hard harvest on the Bering Sea. *National Geographic* 182(4):72-103.

Ianelli, J.N. and Wespestad, V. (1998). Trends in North Pacific Cod and Pollack Catch 1981-1998. 31 October 1998. Alaska Fisheries Science Center. http://www.refm.noaa.gov/stocks/Presentations/norfish/gfish98.htm

Kachina, T.F. (1981). *Herring Population in the Western Bering Sea.* Light Industry, Moscow. 121 pp.

McCrae, J. (1994). Scarlet king crab - *Lithodes couesi*, Oregon hair crab - *Paralomis multispina*. Oregon Developmental Series. Oregon Department of Fish and Game. Viewed 31 October 2000. http://www.hmsc.orst.edu/odfw/devfish/sp/crab.html.

Mitin, S. (1997). Border patrol boats will protect fish stocks. *Izvestia,* 8 October 1997, p. 2. In *The Current Digest of the Post-Soviet Press* 49(40):14-15.

Monakhov, V. (1998). The fishery industry in the Russian Far East. *FAO/EASTFISH Fishery Industry Profile* 19. FAO, Copenhagen. Pp. 74.

Myles, K. (2000). President Putin criticises fishing industry. *FIS Hot News*, 27 July 2000. http://fis.com/fis/hotnews/.

Naumenko, N.N. and Bonk, A.A. (1999a). Herring of Korf-Karaginsky (Resources and Fisheries). *Rybnoye Khozyaistvo* 1: 27-28.

Naumenko, N.N. and Bonk, A.A. (1999b). On Advisability of Limited Commercial Use of Spawning Korfo-Karaginsky Herring. Issues of Protection and Sustainable Use of Kamchatka's Biological Resources. Theses reports. Petropavlovsk-Kamchatsky.

Pautzke, C.G. (1997). *Russian Far East Fisheries Management.* North Pacific Fishery Management Council Report to Congress, 30 September 1997. North Pacific Fishery Management Council, Anchorage, Alaska.

Plotnikov, N. (1997). Was Boris Yeltsin right? - Attempts to protect national interests encounter stiff resistance. *Nezavisimaya Gazeta*, 23 October 1997, p. 3. In *The Current Digest of the Post-Soviet Press* 49(43) :13-14.

Reznik, B. and N. Ostrovskaya. (1994). Operation fishing season: protecting our fish on the border - homegrown poachers still being caught. *Izvestiya*, 30 April 1994, p. 2. In *Foreign Broadcast Information Service-SOV* 86: 11-12.

Ruchkin, A. (1997). The war, as always, starts at the border - the war against crime. *Rossiiskaya Gazeta*, 7 October 1997, p. 2. In *The Current Digest of the Post-Soviet Press* 49(40):15.

Schulte-Paul, I. (1999). A thousand-year-old fish product on its way to the future: An overview of the *surimi* market. *Eurofish* 2/99: 74-78.

Shuntov, V.A. and Dulepova, E.P. (1995). Current status and biological and fish productivity of the Bering Sea ecosystem. Pp. 358-387 in *Complex Studies of the Bering Sea Ecosystem: Compilation of Scientific Works*. VNIRO Publishers, Moscow.

Sviridova, Y. (1998). Neither big, nor small fish rush to Russian nets. *Northern Pacifica* 1(5).

Watson, L.J. (1994). Shrimp. ADF&G Wildlife Notebook Series. Alaska Department of Fish and Game. http://www.state.ak.us/adfg/notebook/shellfsh/shrimp.htm. Viewed 31 October 2000.

Witherell, D. (2000). *Groundfish of the Bering Sea and Aleutian Islands Area: Species Profiles 2000.* North Pacific Fishery Management Council, Anchorage, Alaska. Pp. 101.

Zaitsev, G. (1996). The fishery industries in Russia. *FAO/GLOBEFISH Research Programme* 43. FAO, Rome. Pp. 41.

Zilanov, V.K. (1996). Commercial fishing in Russia: Stabilization of stagnation? *Sakhalin Fisherman* 42(96), 18 October 1996. Translated from Russian and published in Anon., (1999). *The 1999/2000 Directory: Russian Far East Fishing Companies.* Pacific Rim Fisheries Program, Institute of the North, Alaska Pacific University, Anchorage, Alaska. Pp. 135-141.

Annex I

Institutions and terminology associated with Bering Sea fisheries, in English and Russian

Act on Fishery Regulation
Правила рыболовства (*Pravila Rybolovstva*)

Electronic fishing log
электронный прмысловый журнал (*Electronnyi Promyslovyi Zhurnal*)

External Intelligence Service
СВР (*SVR (Sluzhba Vneshney Razvedki)*)

Federal Border Sevice
Федеральная пограничная служба (ФПС) (*Federal'naya pogranichnaya sluzhba (FPS)*)

Federal Security Service
ФСБ (*FSB, Federal'naya Sluzhba Bezopasnosty*)

Federal Service of Tax Police
ФСНП (*FSNP, Federal'naya Sluzhba Nalogovoy Politzii*)

Fisheries Department of Ministry of Agriculture
Департамент по рыболовству (*Departament po Rybolovstvu*, temporary name that was reverted to the former name, *Goscomrybolovstvo*)

«Fisheries» information system
Информационная система «Рыболовство» (*Informatzionnaya Systema «Rybolovstvo» (ISR)*)

«Kamchatka Fisherman» newspaper
«Рыбак Камчатки» («*Rybak Kamchatky*»)

Kamchatka Regional Directorate for Protection and Reproduction of Fish Stocks and Regulation of Fishing
Камчатрыбвод (КРВ) (*Kamchatrybvod (KRV)*)

Ministry of Foreign Affairs
МИД (*MID, Ministerstvo Inostrannykh Del*)

Ministry of Interior
МВД (*MDV, Ministerstvo Vnutrennikh Del*)

Regional Committee on Enironmental Protection
Областной комитет по охране природы (*Oblastnoy Commitet po Okhrane Okruzhayushey Sredy (Oblcompriroda)*)

Regional Directorate of Federal Security Service
Управление ФСБ (УФСБ) (*Upravlenie FSB (UFSB)*)

Regional Directorate of Interior
Управление внутренних дел (*Upravlenue Vnutrennikh Del (UVD)*)

Regional Fisheries Inspection
Управление рыбоохраны (*Upravlenie rybookhrany*)

Regional Inspectorate for protection of marine biological resources
Региональная инспекция охраны морских биологических ресурсов (РИОМБР) (*Regional'naya Inspektzia Okhrany Morskikh Biologicheskikh Resursov (RIOMBR)*)

Russian Foreign Exchange and Export Control Service
служба России по валютно-экспортному контролю (*Sluzhba Rossii po Val'utno-exportnomu Kontrol'u*)

Russian Gazette
Rossijskaya Gazeta

Special Marine Inspectorate
Специальная морская инспекция (Spetzial'naya Morskaya Inspektzia – Spetzmorinspectzia)

State Committee for Statistics
Госкомстат (*Goscomstat*)

State Committee for Customs
ГТК (*GTK, Gosudarstvennyi Tamozhennyi Commitet*)

State Directorate for Protection and Reproduction of Fish Stocks and Regulation of Fishing
Главрыбвод (*Glavrybvod*)

State Committee for Fisheries
Госкомрыболовство (*Goscomrybolovstvo*)

State Taxation Service
ГНС (*GNS, Gosudarstvennaya Nalogovaya Sluzhba*)

TINRO
Pacific Research Institute of Fisheries and Oceanography

Total Allowable Catch
ОДУ (*ODU, Obshiy Dopustimyi Ulov*)

VNIRO
Russian Federal Research Institute of Fisheries and Oceanography

Annex 2

Reported catches of major industrial fisheries in the North Pacific

The Bering Sea partly covers two of the Major Fishing Areas defined by FAO for statistical purposes: the Pacific Northwest (Area No. 61) (see **Figure 2** in main report) and the Pacific Northeast (Area No. 67). The data presented in the following tables and graphs show annual catch amounts from these two areas, reported to FAO by the responsible government agencies of the fishing nations concerned.

Source: FAO Fishstat, January 2000.

Note: Where a dash or blank is shown in the tables, this usually means "zero catch", i.e. that no catch was reported to FAO: in some places a zero catch is clearly inconsistent with the trend over previous and/or following years.

Table 1

Alaska Pollack catch (1000 t) in the Pacific Northeast (NE) and Northwest (NW), 1984-99

Country	1984	1985	1986	1987	1988	1989	1990	1991	1992	1993	1994	1995	1996	1997	1998	1999
Canada (Pacific NE)	0.6	1.7	0.6	0.9	0.5	0.4	0.7	2.6	3.2	8.1	4.7	3.3	2.1	1.9	0.9	1.2
China (Pacific NE)	0.0	0.0	0.0	20.0	20.0	20.0	20.0	30.0	40.0	40.0	40.0	50.0	45.0	40.0	50.0	40
Germany (Pacific NE)	23.8	0.0	0.0	0.0	0.0	0.0	0.0	0.0	0.0	0.0	0.0	0.0	0.0	0.0	0.0	
Japan (Pacific NE)	512.3	379.8	169.4	96.8	0.0	0.0	0.0	0.0	0.0	0.0	0.0	0.0	0.0	0.0	0.0	
Rep. of Korea (Pacific NE)	315.4	329.1	463.7	288.6	132.9	114.4	76.2	44.2	81.0	44.8	14.6	11.0	2.5	0.0	0.0	
Poland (Pacific NE)	53.7	30.8	6.9	0.0	0.0	0.0	0.0	0.0	0.0	0.0	0.0	0.0	0.0	0.0	0.0	
USSR / Rus. Fed. (Pacific NE)	11.7	1.4	1.4	0.5	0.1	0.2	0.0	0.3	0.0	0.0	0.0	0.0	0.3	0.0	0.0	
USA (Pacific NE)	455.1	656.4	963.3	1307.7	1396.8	1359.1	1432.2	1295.2	1339.1	1477.8	1417.3	1293.9	1189.8	1139.6	1232.2	1055.0
Subtotal (Pacific NE)	**1372.7**	**1399.2**	**1605.2**	**1714.5**	**1550.3**	**1494.2**	**1529.1**	**1372.2**	**1463.4**	**1570.8**	**1476.6**	**1358.2**	**1239.7**	**1181.5**	**1283.1**	**1096.2**
China (Pacific NW)	0.0	0.0	0.0	0.0	0.0	0.0	0.0	91.3	131.7	135.0	130.2	199.5	181.9	180.0	141.4	116.7
Japan (Pacific NW)	1092.6	1153.3	1252.4	1215.7	1259.1	1153.8	871.4	540.9	498.8	382.3	379.3	338.5	331.2	338.8	316.0	382.4
DPR Korea (Pacific NW)	0.0	0.0	0.0	0.0	0.0	0.0	0.0	0.0	0.0	0.0	75.1	120.2	15.4	66.6	60.0	55.0
Rep. of Korea (Pacific NW)	83.2	122.3	155.3	142.2	179.8	271.0	245.3	153.6	254.4	181.3	296.9	334.9	224.4	223.1	236.3	146.2
Poland (Pacific NW)	0.0	115.9	163.2	230.3	298.7	268.6	223.5	230.6	297.7	235.2	270.0	249.3	116.3	125.4		65.5
USSR / Rus. Fed. (Pacific NW)	3437.9	3341.6	3582.8	3421.2	3369.8	3133.0	2863.9	2495.6	2340.7	2114.5	1746.6	2208.4	2439.7	2252.7	1930.7	1500.5
Taiwan Pro.China (Pacific NW)	0.0	0.0	0.0	0.0	1.0	0.5	0.0	0.1	0.0	0.0	0.0	0.0	0.0	0.0	0.0	9.0
Ukraine (Pacific NW)	0.0	0.0	0.0	0.0	0.0	0.0	3.0	9.2	0.0	0.0	0.0	0.0	0.0	0.0		
Subtotal (Pacific NW)	**4613.6**	**4733.1**	**5153.8**	**5009.5**	**5108.3**	**4826.8**	**4207.1**	**3521.3**	**3523.3**	**3048.2**	**2898.1**	**3450.8**	**3308.8**	**3186.6**	**2684.4**	**2275.2**
TOTAL	**5986.3**	**6132.3**	**6758.9**	**6724.0**	**6658.6**	**6321.0**	**5736.2**	**4893.5**	**4986.7**	**4619.0**	**4374.8**	**4809.0**	**4548.5**	**4368.1**	**3967.4**	**3371.4**

Figure 1
Alaska Pollack catch (1000 t) in the Pacific Northwest, 1984-99

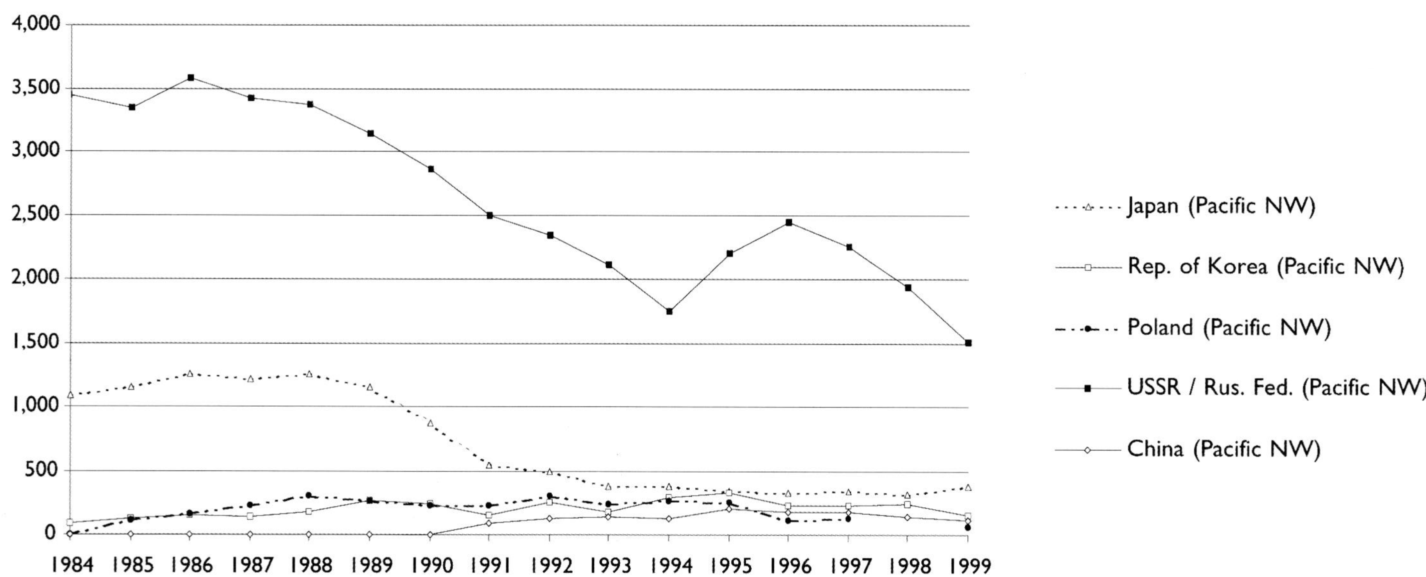

Table 2 and Figure 2

King crab catch (t) in the Pacific Northeast (NE) and Northwest (NW), 1984-99

Country	1984	1985	1986	1987	1988	1989	1990	1991	1992	1993	1994	1995	1996	1997	1998	1999
USA (Pacific NE)	7804	6969	11752	13184	9513	11971	15385	12764	8644	11218	5425	6656	9526	8177	10941	7675
Japan (Pacific NW)	137	351	123	207	329	623	824	1122	2058	313	472	261	322	154	132	117
Rep. of Korea (Pacific NW)	6	14	5	4	10	3	3	2	11	9						
TOTAL	**7947**	**7334**	**11880**	**13395**	**9852**	**12597**	**16212**	**13888**	**10713**	**11540**	**5897**	**6916**	**9848**	**8331**	**11073**	**7792**

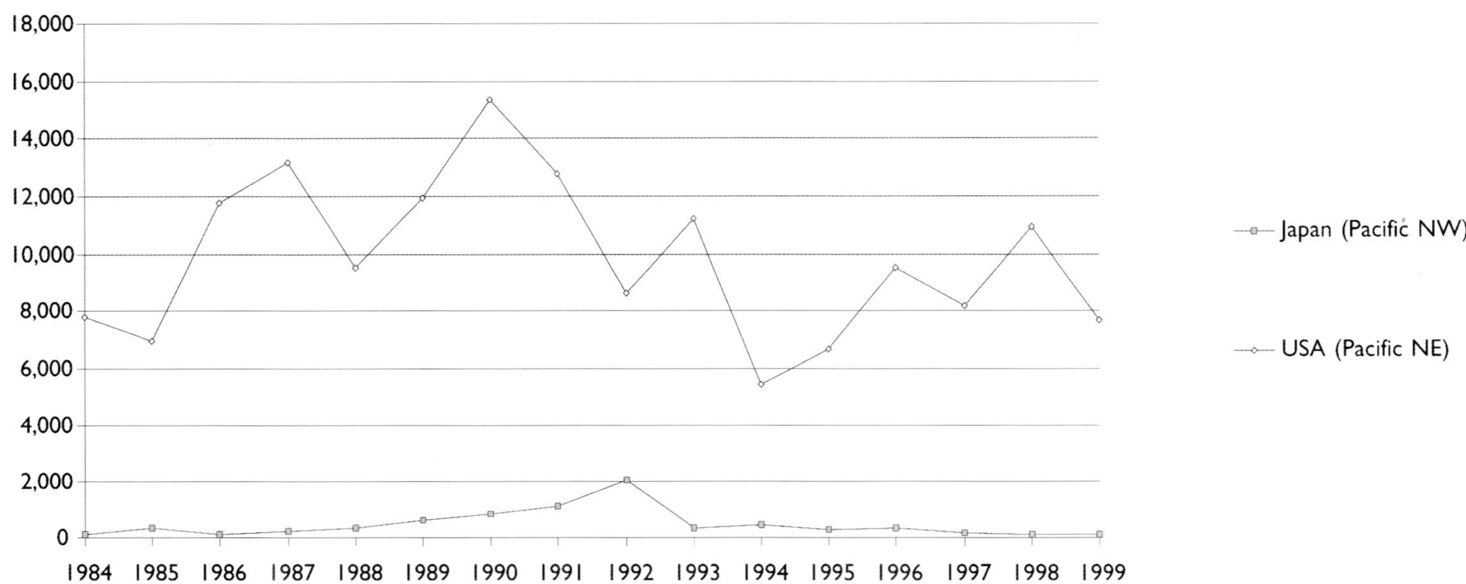

Table 3

Pacific Cod catch (t) in the Pacific Northeast (NE) and Northwest (NW)

Country	1984	1985	1986	1987	1988	1989	1990	1991	1992	1993	1994	1995	1996	1997	1998	1999
Canada (Pacific NE)	3459	2345	3668	13691	11024	9143	6233	11910	10125	8123	3550	2129	693	1537	1354	835
Japan (Pacific NE)	44119	38255	43027	54916	0	0	0	0	0	200	220	0	0	0	0	0
Rep. of Korea (Pac. NE)	11429	6362	11842	2654	653	0	0	0	0	0	0	0	0	0	0	0
USSR/Russ. Fed (Pac. NE)	345	84	20	0	286	6	0	2	0	0	0	0	179	0	0	0
USA (Pacific NE)	90950	90929	113390	136900	232729	213224	254927	251164	249717	218996	208785	268257	274569	299970	252197	237679
Subtotal (Pacific NE)	**150302**	**137975**	**171947**	**208161**	**244692**	**222373**	**261160**	**263076**	**259842**	**227319**	**212555**	**270386**	**275441**	**301507**	**253551**	**238514**
Japan (Pacific NW)	69861	79296	57481	56838	59100	57689	59085	48925	75690	62286	65780	56561	57576	58477	57243	55292
Rep. of Korea (Pac.NW)	953	3936	1579	1508	1206	3030	490	1387	3966	11626	3702	2476	2740	3984	6249	6509
Russ. Fed. (Pac. NW)	140178	188359	168425	175271	148111	140359	121397	106517	154297	95823	81445	100730	93711	79927	94282	101929
Subtotal (Pacific NW)	**210992**	**271591**	**227485**	**233617**	**208417**	**201078**	**180972**	**156829**	**233953**	**169735**	**150927**	**159767**	**154027**	**142388**	**157774**	**163730**
TOTAL	**361410**	**409593**	**399437**	**441778**	**453109**	**423451**	**442132**	**419905**	**493795**	**397054**	**363482**	**430153**	**429468**	**443895**	**411325**	**402244**

Table 4

Saffron Cod catch (t) in the Pacific Northwest (NW)

Country	1984	1985	1986	1987	1988	1989	1990	1991	1992	1993	1994	1995	1996	1997	1998	1999
Russ. Fed. (Pac. NW)	34879	22613	26522	27929	26954	27841	32280	29918	50624	43079	13213	25566	21110	27803	40426	47032

Figure 3

Pacific Cod catch (t) in the Pacific Northwest (NW), 1984-99

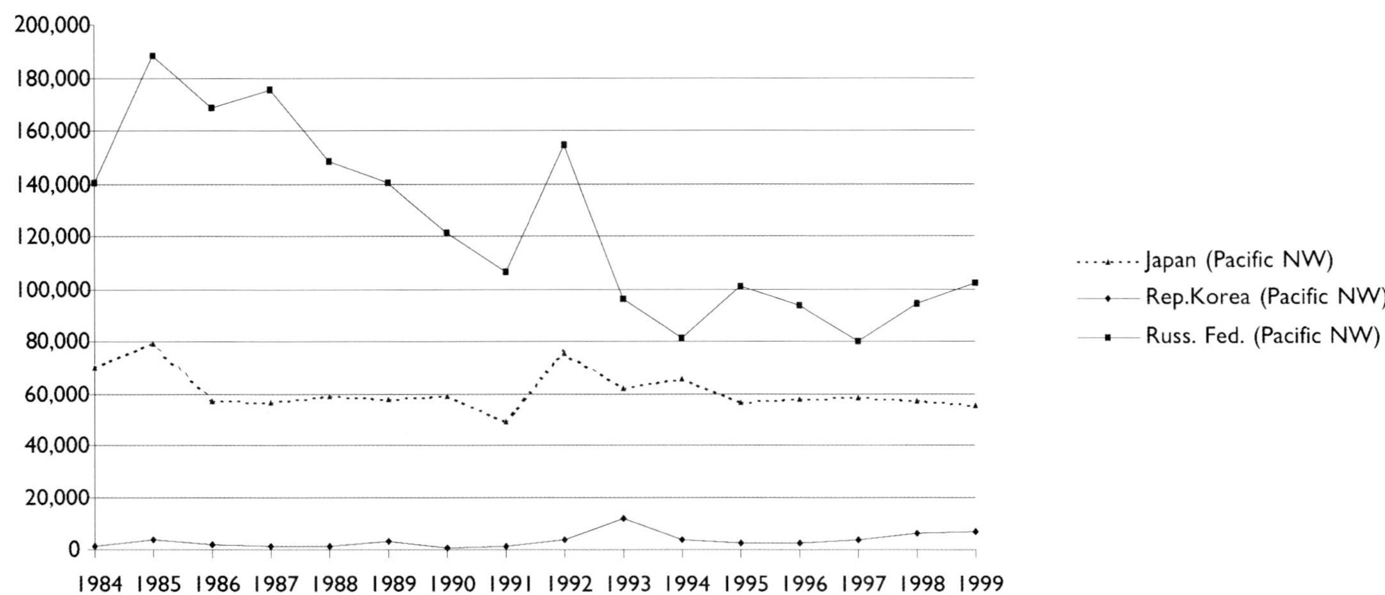

Table 5 and Figure 4

Pacific Herring catch (t) in the the Pacific Northeast (NE) and Northwest (NW), 1984-99

Country	1984	1985	1986	1987	1988	1989	1990	1991	1992	1993	1994	1995	1996	1997	1998	1999
Canada (Pacific NE)	33703	25955	16491	37809	31601	41008	41280	40030	34867	41321	40909	27036	22146	31585	23300	29400
USA (Pacific NE)	43730	56549	51764	47255	52690	44936	41698	48606	66920	44492	48211	48666	49300	52892	39878	39038
Subtotal (Pacific NE)	**77433**	**82504**	**68255**	**85064**	**84291**	**85944**	**82978**	**88636**	**101787**	**85813**	**89120**	**75702**	**71446**	**84477**	**63178**	**68438**
China (Pacific NW)	9444	3228	6445	12971	12026	5456	3941	2472	2209	994	1330	2325	1665	15780	21796	17936
Japan (Pacific NW)	6760	9193	72741	18652	5855	5890	1550	13837	2854	1482	2146	3873	2021	1926	2531	2579
Rep. of Korea (Pacific NW)	497	773	1386	2763	6075	6697	6317	5001	6079	3983	3865	8622	5525	13214	13340	21446
USSR/Russ. Fed (Pac. NW)	150161	189584	168356	220712	189477	99966	97357	98326	109279	115148	85218	116787	171810	313397	395595	359194
Subtotal (Pacific NW)	**166862**	**202778**	**248928**	**255098**	**213433**	**118009**	**109165**	**119636**	**120421**	**121607**	**92559**	**131607**	**181021**	**344317**	**433262**	**401155**
TOTAL	**244295**	**285282**	**317183**	**340162**	**297724**	**203953**	**192143**	**208272**	**222208**	**207420**	**181679**	**207309**	**252467**	**428794**	**496440**	**469593**

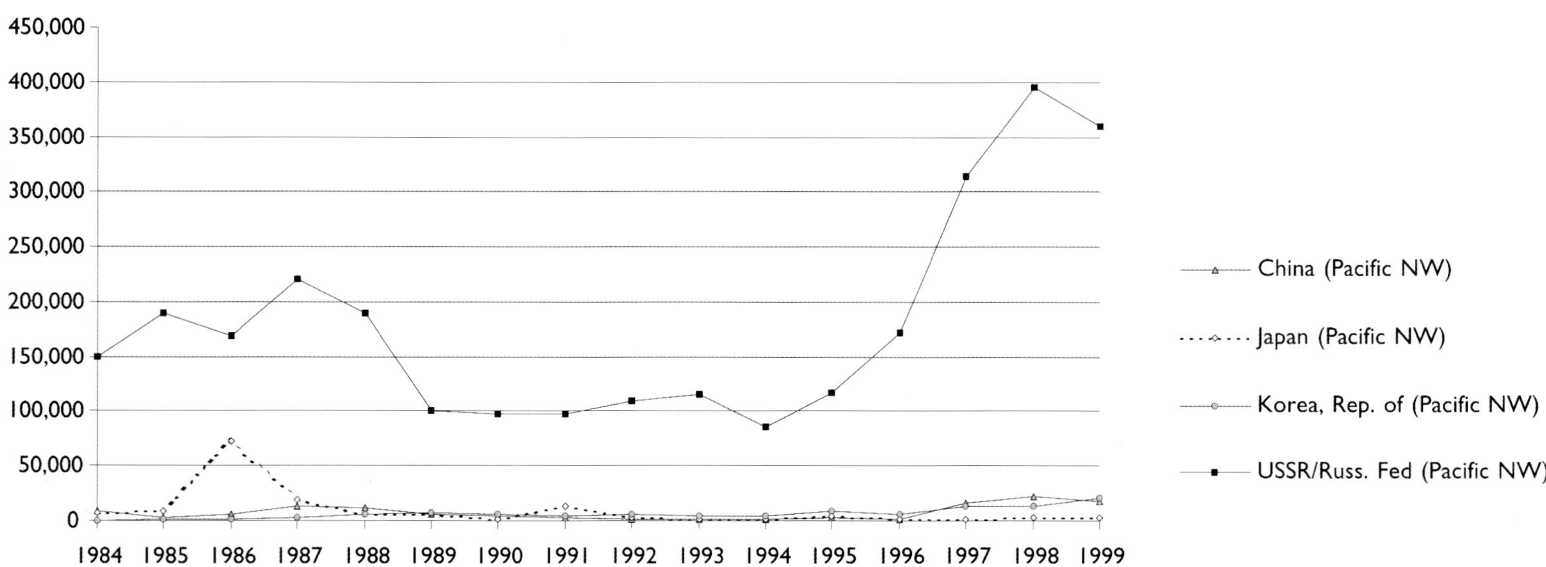

Table 6 and Figure 5

Pacific Ocean Perch catch (t) in the Pacific Northeast (NE) and Northwest (NW), 1984-99

Country	1984	1985	1986	1987	1988	1989	1990	1991	1992	1993	1994	1995	1996	1997	1998	1999
Canada (Pacific NE)					6789	6068	5598	4295	3990	4609	5649	5103	6174	5782	6184	6900
Germany (Pacific NE)	2	0	0	0	0	0	0	0	0	0	0	0	0	0	0	-
Japan (Pacific NE)	3584	3	13	4	238	0	0	0	0	0	0	0	0	0	0	-
Poland (Pacific NE)	47	9	1	69	82	129	5	1	0	0	0	0	0	0	0	-
Russian Fed. (Pacific NE)	6	0.5	61	282	157	115	0	0	0	0	0	0	0	0	0	1614
USA (Pacific NE)	5973	4378	7342	9620	5406	10130	27657	13605	20630	20649	14000	15613	21004	19580	18445	20608
Subtotal Pacific Northeast	**9612**	**4391**	**7417**	**9975**	**12672**	**16442**	**33260**	**17901**	**24620**	**25258**	**19649**	**20716**	**27178**	**25362**	**24629**	**31121**
Japan (Pacific NW)	6757	3776	1992	2011	2452	1811	905	852	1621	2760	1520	1449	1730	1334	896	766
Rep. of Korea (Pacific NW)	0	0	0	0	0	0	0	0	4	0	0	0	0	0	0	-
Russian Fed. (Pacific NW)	151	412	240	627	930	538	334	209	560	1153	3447	3539	2615	2095	1544	864
Subtotal Pacific Northwest	**6908**	**4188**	**2232**	**2638**	**3382**	**2349**	**1239**	**1061**	**2185**	**3913**	**4967**	**4988**	**4345**	**3429**	**2440**	**1630**
TOTAL	**16520**	**8578**	**9649**	**12613**	**16054**	**18791**	**34499**	**18962**	**26805**	**29171**	**24616**	**25704**	**31523**	**57582**	**54138**	**30752**

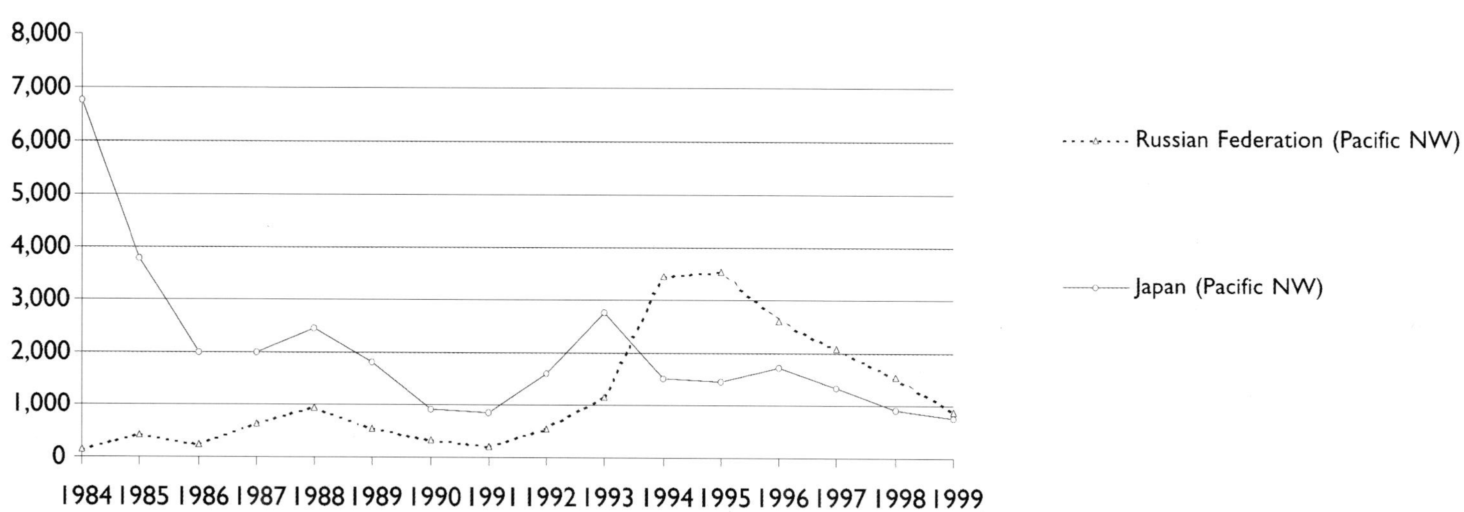

Table 7 and Figure 6

Pandalus shrimp catch (t) in the Pacific Northwest (NW) (t), 1984-99

Country	1984	1985	1986	1987	1988	1989	1990	1991	1992	1993	1994	1995	1996	1997	1998	1999
USSR/Russ. Fed. (Pacific NW)	905	1265	1065	4411	508	1478	1326	1609	2028	2462	1989	2398	3000	2426	4914	-
TOTAL	**905**	**1265**	**1065**	**4411**	**508**	**1478**	**1326**	**1609**	**2028**	**2462**	**1989**	**2398**	**3000**	**2426**	**4914**	**-**

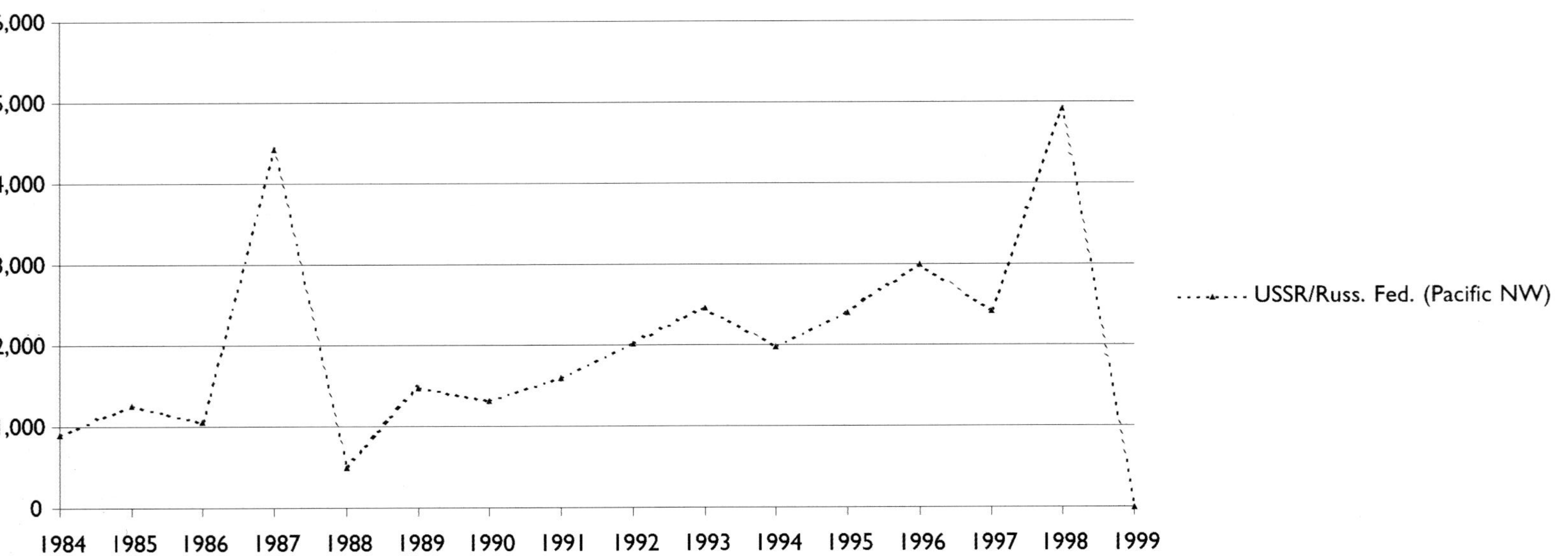